SAFE BY ACCIDENT?

SAFE

BY ACCIDENT?

Take the LUCK out of SAFETY
Leadership Practices that Build a Sustainable Safety Culture

JUDY L. AGNEW
AUBREY C. DANIELS

Performance Management Publications (PMP)

Performance Management Publications (PMP)
3344 Peachtree Road NE, Suite 1050
Atlanta, GA 30326
678.904.6140
www.PManagementPubs.com

ISBN: 978-0-937100-18-9

2 3 4 5 6 7

Cover and text design: Lisa Smith
Editor: Gail Snyder
Production Coordinator: Laura-Lee Glass
Set in Adobe Garamond Pro

PMP books are available at special discounts for bulk purchases by
corporations, institutions, and other organizations. For more information,
please call 678.904.6140, ext. 131 or e-mail lglass@aubreydaniels.com

For Matthew and Kianna—
my greatest source of reinforcement.

J.A.

Acknowledgements

As always we are indebted to the field of behavior analysis and to our clients in equal measure. Behavior analysis provides the conceptual framework to understand organizational behavior and our clients help us continually refine and improve the application of the science.

We both feel fortunate to work with an extraordinary team of consultants at ADI whose hard work and dedication to finding the best solutions for our clients provided the foundation for this book. Many of those directly participated in bringing this book to fruition. Cloyd Hyten's and Bart Sevin's commitment to improving our safety business served as the impetus for getting started. Cloyd Hyten, Darnell Lattal, and Gerald Perrier read the manuscript and provided helpful comments that improved the book in every case. Gail Snyder, with her understanding of behavior analysis and exceptional writing skills, provided invaluable editing. Laura-Lee Glass, Julie Terling, and the people at PKPR helped ensure the book appealed to the target audience. Laura-Lee Glass handled the production of the book with great skill and good humor. And Lisa Smith did her usual magic with the cover and layout.

No one can whistle a symphony.
It takes an orchestra to play it.
– H.E. Luccock

Contents

Luck never gives; it only lends.
— Swedish proverb

Are You Gambling with Safety?

Time after time, BP appeared to have gambled with safety.
– Capt. Hung Nguyen,
Co-chair of the Deepwater Horizon Joint Investigation team

Despite decades of reduction in safety related deaths and injuries on-the-job, catastrophic accidents appear to be on the rise. In the last ten years, there have been at least nine industrial accidents that resulted in more than 125 deaths. Two accidents in 2010, the BP oil spill in the Gulf of Mexico and the Upper Big Branch mining disaster in West Virginia, have not only caused injury and death; they have wreaked economic havoc on the communities in which they occurred, and incalculable environmental impact in the case of the gulf disaster.

Why are catastrophic accidents on the rise?

BP's Deepwater Horizon offshore oil rig operated for seven consecutive years without a single lost-time incident or major environmental event. By that measure they were safe. We now know that for those seven years they were "safe by accident." By all reports, many unsafe conditions and behaviors existed at all levels of the organization that had, through sheer luck, not resulted in an accident. On April 20, 2010, the luck ran out.

This pattern is not unique to the Deepwater Horizon case. Based on decades of research and work with many of the world's leading corporations, we've concluded that many companies are safe by accident because they focus too heavily on incident rate and don't take a scientific approach to managing safe and at-risk behavior. Sophisticated companies that use only the latest scientific information and technologies from chemistry, physics, engineering, and biology, use so-called common sense, myth, and downright faulty information to manage the behavior of their employees.

In this book, we will reveal how an in-depth knowledge of the science of behavior can enable leaders and safety professionals to build systems and management practices that create a lasting corporate-wide commitment to safety—from the boardroom to supervisors to the front lines. Organizations that fail to take a scientific approach to safety's human-behavior element are gambling with their futures and are ultimately only safe by accident.

The Safety Leader's Role Has Been Poorly Defined

This is not intended to be an indictment of leaders, because the role of leaders in safety has been poorly defined. Vague phrases such as "making safety a priority" or "creating a safety culture" have little meaning. What should a leader do today and tomorrow to ensure a safe work environment?

After interacting with thousands of supervisors, managers, and executives, we have rarely met a leader who didn't care about safety. Lack of caring and concern about safety is not the problem. Leaders are adept at talking the talk: "Safety is the first priority," "Nothing is more important than safety!" Nevertheless, when we ask leaders about the activities they do each day regarding safety, we often hear general phrases like "I make sure the employees know how important safety is" or "I emphasize safety all the time."

When they do get more specific, we hear things like, "I remind them to wear their PPE" or "I talk about safety each day" or "I start every meeting with safety."

Two questions come to mind: (1) Are these the right behaviors for leaders to engage in? and (2) Is it enough? Many leaders we work with have a nagging feeling the answer to both questions is no. Nevertheless, they aren't sure what more to do. How do you become an exemplary safety leader? What can a leader do to help create a culture that truly embraces safety? What are the steps to build a high-performance safety culture?

The first step is to define *culture* in a way that enables and directs action. Most definitions of *culture* are values-based ("We are committed to doing business in a manner that protects our employees and the environment"). While this is an excellent starting point, such definitions do not clarify how to make it happen. We define *culture* this way:

> Patterns of behavior (what we say and do),
> encouraged or discouraged, inadvertently or intentionally,
> by people or systems over time.

Each component of this definition is explained below.

PATTERNS OF BEHAVIOR (WHAT WE SAY AND DO)
Culture is expressed through behavior. As noted above, organizations typically describe the desired safety culture in terms of values. Values are the foundation but the culture as it exists today is expressed through what is said and done. Having good safety values is only helpful if those values translate into behavior.

ENCOURAGED OR DISCOURAGED
Since culture is expressed through patterns of behavior, and behavior is influenced by consequences, it follows that those patterns of behavior are influenced by consequences. Good patterns of behavior (those consistent with safety values) can be encouraged or

discouraged (reinforced or punished). Poor patterns of behavior (those inconsistent with safety values) can also be encouraged or discouraged. Culture must be carefully and purposefully cultivated.

INADVERTENTLY OR INTENTIONALLY

Because so many different consequences operate within organizations, quite frequently the wrong behaviors are inadvertently (accidentally) reinforced. For example, skipping steps of a lockout/tag-out procedure gets reinforced when the mechanic repairs equipment faster and gets production running again. Alternatively, the right behaviors often get punished (discouraged). For example, an overly cumbersome reporting system inadvertently discourages near-miss reporting.

BY PEOPLE OR SYSTEMS OVER TIME

Behavior is encouraged and discouraged by people (executives, management, peers, customers, et cetera) and systems (pay, promotion, incentives, communication, procedures, and so on). Creating a high-performance safety culture requires evaluating all possible sources of behavioral consequences and ensuring that those consequences encourage desired safe behavior.

Good Intentions Are Not Enough

Safety cultures must be carefully and deliberately nurtured. If not, they will develop inadvertently and, more often than not, they will drift toward poor safety habits. Ineffective safety cultures can develop despite good intentions. In our experience, management almost always has good intentions, but intention does not always match impact. Good safety leadership requires systematic assessment of the impact of management actions:

- Do the safety programs deliver the desired outcomes?

- Are the communication systems effective at disseminating information?

- Do individual leaders effectively coach direct reports in safety?

In sum, do the safety efforts of management result in improved safety? One of the goals of this book is to share some common safety management practices that are used with the best of intentions, but that actually work against the development of a high-performance safety culture.

Creating a high-performance safety culture requires

- identification of the desired patterns of behavior,

- analysis of how those patterns of behavior are currently encouraged or discouraged by people or systems, and

- alignment of the people and systems so they intentionally and systematically encourage the desired patterns of behavior.

This is no simple task. The good news is that you are doing some—perhaps many of these things—already. However, some of our recommendations will be new. We are also going to recommend you stop doing some things that you spend time and money on that won't lead to a high-performance safety culture. We are then going to recommend that you do other things that will have a positive impact.

While step-by-step instructions are not possible given each organization is at a different point on the road to safety excellence, the best tool we can give you is a set of guiding principles: the science of behavior analysis. We will also share seven common safety leadership practices that don't work and make recommendations on what to do instead. Ultimately, we hope to provide a framework of activities that will help move your organization toward a high-performance safety culture. We hope you agree that leaders don't need another lofty book about "safety vision" and "safety commitment."

Instead, leaders need applicable information and instruction. Our goal is to help safety leaders take action, know how to assess the impact of that action, and adjust future action if necessary. Rather than relying on luck or wishful thinking, our goal is to help you build a purpose-driven, high-performance safety culture.

Important Notes About Terminology

Safety Leader

While anyone can be a leader in safety, this book is written specifically for those in formal leadership positions: supervisors, managers, and particularly senior leaders who have broader spans of control. While senior leaders have the decision-making power to make many of the changes we recommend, it is important for all those who manage others to show good safety leadership at their own level as well as to reinforce the good safety leadership of those above them on the corporate ladder.

Leadership and Management

In the book *Measure of a Leader*,[1] Aubrey Daniels and James Daniels make the following distinction: managers are given authority by virtue of their position in the organization. They can hire, fire, promote, demote, and change the work assignment of employees. A leader, on the other hand, gains authority when the followers willingly give discretionary effort to the leader's cause. Without followers there are no leaders. Without willing followers there are no effective leaders. Both management and leadership skills are required in safety. Ideally, every supervisor, manager, and executive should possess both skill sets. Safety processes require highly skilled management to develop employees who follow the processes without error or hesitation. Leaders are required to anticipate changes in processes, equipment, and physical skills that will make existing processes more effective.

This book describes how to engineer consequences to sustain safe behavior. Some consequences come from management-owned systems such as incentives, discipline, and the physical work environment. But to attain safety excellence, everyone must be engaged in the safety process. Engagement is the responsibility of leadership. Engagement requires good leadership because engagement is generated and supported by follower-granted consequences. Because both good *leadership* and good *management* are required for a high-performance safety culture, we will use the terms interchangeably throughout the book.

Safe Behavior

We use the phrase *safe behavior* to refer to any behavior that has a positive impact on safety outcomes. The phrase is most often used to refer to the actions of frontline employees (actions such as putting on PPE, following safety procedures, keeping walkways clear, and so on), but supervisors, managers, and executives also engage in behaviors that either support or hinder safety. For example, they support or hinder safety by the issues they talk about in meetings, by the behaviors and results they choose to reinforce and choose to ignore, and by the decisions they make regarding equipment purchases or process changes. While these behaviors have an indirect impact on safety, the impact can be profound. Thus, we label behavior at all levels as *safe* or *at-risk*. **When we talk about encouraging safe behavior, we are talking always about behavior at all levels of the organization.**

Behavior-Based Safety (BBS)

Behavior-based safety (BBS) has come to mean a frontline-driven process of targeting particular safe behaviors, arranging peer observations, and providing feedback and reinforcement. In truth, all good BBS processes have a heavy management component, but this fact is often not well publicized. This book is not about BBS per se. It is about a behavioral approach to safety leadership. The ideas and technology presented in this book can be used in conjunction with a behavior-based safety process or as a stand-alone process for safety leaders. In some cases there is great value in working on safety leadership prior to implementing BBS. When the management/leadership components of safety are functioning well, a BBS implementation will be much more effective, the willingness of the hourly population to participate will be higher, managers and leaders will be primed to play their part, and fewer roadblocks to improvement will arise.

---

[1]Daniels, A.C. & Daniels, J.E. (2007) *Measure of a Leader: An Actionable Formula for Legendary Leadership.* New York: McGraw-Hill, Inc.

PART ONE

The Science of Behavior

CHAPTER 1

The Science of Behavior

The ABC Model and
the Role of Consequences

The older I get, the less I listen to what people say
and the more I look at what they do.
– Andrew Carnegie

Consider the following:

- Airport security screeners consistently miss banned objects.

- Security guards have been caught sleeping.

- Pilots observed skipping steps during "rolling checks."

- Hospital-borne infection rates continue to climb when hand washing will solve the problem.

- A company is having difficulty sustaining a Lean-Sigma initiative.

- A black belt, six-sigma is doing a project to "turn rail cars" more efficiently—the last project resulted in a $6 million accident.

- A company has a serious morale problem due to a recent layoff.

- Employees fail to take responsibility for errors.

While these problems come from a wide range of businesses and activities, they have two things in common. First, they represent a problem of safety leadership. Firing, disciplining, or lecturing frontline employees will not fix any of these problems. In fact, those actions by management may even make such situations worse. Second, they can all be solved with a better understanding of behavior analysis.

What Is Behavior Analysis?

Behavior analysis is the scientific study of behavior. Applied behavior analysis is the application of the science to bring about positive change in socially significant behavior.[1] Behavior analysis is the foundation science of all good behavior-based safety processes.[2] Behavior-based safety (BBS) is the application of principles of behavior analysis to bring about positive change in behaviors that will promote safety within work settings. As noted earlier, good BBS processes include changing the behavior of management in order to facilitate working safely at the front line.

Behavior analysis helps you understand choice, decision making, and the effects of performance on results. It helps you understand the thoughts and feelings that can interfere with making safety the top priority when demands of the business are screaming for attention. Behavior analysis can help you accomplish strategic imperatives while keeping safety as a screen through which all organizational activities must pass. The science provides a lens that allows you to see the complexity of behavior while helping you forge a path to a high-performance safety culture.

This book is about how to improve safety leadership through the application of behavior analysis. Our position in this book can be best summarized by the poet, Alexander Pope, who said, "Some people will never learn anything, for this reason, because they understand everything too soon." Too many people who have been exposed to behavior analysis think they have its essence when they are introduced to concepts like reinforcement, reward, feedback, and measurement. Yet few of them know that thousands of research projects have been conducted for understanding precisely how these and other concepts derived from the science work in real life. In their book, *Behavior Analysis and Learning,* David Pierce and Frank Epling[3] have indexed 49 entries under "reinforcement" and "schedules of reinforcement" alone. Every year at the annual meeting of the Association of Behavior Analysis International (ABAI) over a thousand research studies are presented. In summary, there is much to know about behavior.

Exposing people to the science of behavior analysis has its advantages and disadvantages, but we believe the former outweigh the latter. The advantage is that you don't have to know everything about the science in order to put its findings to good use. Hardly anyone understands the inner technology of computers, for example, but with a little instruction almost everyone can operate one. For those who are taught just enough to do e-mail or type documents, however, the problems begin when the computer doesn't respond to the usual routine. It won't start; it hangs up; it won't load; or it is intolerably slow. At this point, the little knowledge that we have about the computer is not only useless, it is often harmful, because as we keep trying to fix the problem, we sometimes waste countless hours and often do things that unknowingly complicate the repair. In short, we were trained, but not educated. We learn "too soon."

An oft-quoted story illustrates the difference between training and education. A ship is docked and ready for sail but when it is ready to leave, the engines won't start. The engine room crew tries everything they know to no avail. Finally, they summon a local expert to see if he can fix the problem. When he arrives, he asks them to try to start the engines. He listens briefly to some strange and seemingly random noises and asks for a hammer. He walks around looking for a particular pipe. When he finds it, he raps the pipe with a sharp blow and the engine immediately starts. He reaches in his pocket, writes up an invoice and hands it to the Captain.

The Captain looks at it and is startled by the amount. He says to the expert, "You charged us $5,000 to hit a pipe with our hammer!" Without hesitation, the expert responds, "Not at all. I charged you $0.50 to hit the pipe. I charged you $4,999.50 for knowing where to hit it."

The expert knew how the system operated. He was educated in the science behind the operation of steam engines. The crew was trained. They knew some techniques and when those didn't work they were stumped. The difference in that story and behavior-based safety leadership is that if the crew, in their attempts to start the engine, hit the wrong pipe or tried various other techniques, it is most likely that no one would be harmed. On the other hand, when you are dealing with behavior and you do something wrong there are always effects, and most of the time, those effects are bad. Sometimes the effects are even fatal. We want to work with the certain knowledge and confidence that the procedures, processes, and management behaviors we use to manage safety will always work. The only way we know how to do so is by adding a solid, scientific foundation to our workplace repertoire.

Science Versus Technology

Everyone is guilty of being more sensitive to information that supports their position. Science helps us avoid such traps and can correct personal biases. Remember, the goal of the science of behavior is to discover universal laws of behavior. For something to attain the status of a behavioral law, it must apply to all people under all conditions. It would not be correct to say that the laws of behavior wouldn't work in China because it is a Communist country. It might be true that certain behavioral techniques wouldn't work, or be allowed, in a Communist country, but the laws of behavior still work as they have since the beginning of life on earth. Behavioral laws, or principles, are independent of industry, job, organizational position, country, religion, or political persuasion.

While science seeks complete understanding, technology often uses the best available information or data to solve problems. Technology looks for immediate application, but science often says, not so fast. Because technology is based on incomplete information, technologists need to be constantly in touch with the latest scientific discoveries in order to update their practices. At the annual Association of Behavior Analysis International (ABAI) conference, it is difficult to separate the scientific research papers from the technological ones because they are often presented in the same sessions. This is important because old technologies may not only be ineffective when compared to newer ones, but as conditions change, the old technology may even stop working. When scientific research is presented side-by-side, science informs technology. This allows behavioral technology to always be at the forefront of the science.

It is highly unlikely that behavior will stop being a function of its consequences no matter how science progresses. However, new discoveries about consequences may fill gaps in our knowledge that will lead to even more effective technology. Our appeal is that practitioners learn as much as they can about the science in order to reap the many personal and organizational benefits that come from it.

THE A-B-C MODEL

While the ABC Model is elegant in its simplicity, it encompasses the totality of the science. Stated simply, behavior (B) is influenced by what comes before it (A–antecedents) and what follows it (C–consequences). However, as simple as it appears, there are a great many factors determining the effectiveness of each element. Let's look at consequences first.

BEHAVIORAL CONSEQUENCES

The closest thing we have to a law of behavior, as gravity is a law of physics, is "Behavior is a function of its consequences." Behavioral consequences have one of two effects; they either increase or decrease the rate of a behavior. They may increase the rate to some physically maximum rate or they may decrease it to zero.

There are three ways to increase the rate of behavior—**positive reinforcement, negative reinforcement**, and **recovery**.

Positive Reinforcement. No concept is talked about more in safety and understood less than positive reinforcement. It is often defined in terms of words (saying something positive) and tangibles (things that have economic value or cost money). In reality it could be both or neither. Positive reinforcement is defined functionally— by the effect it has on behavior.

Positive reinforcement is any consequence that follows a behavior that increases that behavior's frequency. It is possible that yelling at someone for a safety violation is a positive reinforcer (the recipient enjoys the attention it brings) and as such would increase violations. It is possible that broadcasting the name of a terrorist group claiming responsibility for a bombing is a positive reinforcer, increasing terrorist acts.

The fact that positive reinforcement increases behavior points to the importance of the need to measure behavior in any safety process. If critical safe behavior is not increasing, it is not being reinforced. However, there is a fly in the ointment. Negative reinforcement also increases the rate of behavior.

Negative Reinforcement. If safe behavior is increasing, why would we care whether it is because of positive or negative reinforcement? There are several reasons.

First, negative reinforcement is defined as an increase in the rate of behavior designed to escape or avoid punishment or penalty. Once the punisher or penalty has been avoided or escaped, there is no personal benefit to do more. Thus negative reinforcement leads to sub-optimal performance.

Second, negative reinforcement trades on fear. In other words, people may well follow safety practices, not because they see the personal value in doing so, but because they are afraid of what might happen to them if they don't. Safety cultures driven by negative reinforcement are cultures of fear and distrust where performers are not engaged in safety but are just doing the bare minimum; just enough to stay out of trouble. (Learn more at www.aubrey-danielsblog.com, "Employees have spoken...fear and failed leadership prove disastrous.")

Third, negative reinforcement has detrimental side effects such as suppressed volunteerism, reduced participation, lack of teamwork, covering up mistakes, and (not surprisingly) lower morale.

This do-it-or-else style of managing safety leads to employees behaving safely because they have to; attending safety meetings because they have to; reporting near misses because they have to, and barely meeting the safety goals. This is certainly not by design, but because those in charge of these systems don't have an in-depth understanding of behavioral consequences. They use techniques, processes, and behaviors that unintentionally create a negatively reinforcing safety climate.

Negative reinforcement is always the default approach. What we mean is that in the absence of positive reinforcement, the only consequence left to increase or maintain behavior is negative reinforcement. You don't have to be mean to be a negative reinforcer; you just have to be a manager that doesn't use much social or tangible reinforcement and/or who has not built positive reinforcers into the work process. If you are not actively seeking safe behavior to positively reinforce or trying to find ways to build positive reinforcement into processes, procedures, and policies, you are getting performance by negative reinforcement, like it or not.

The use of negative reinforcement has been the primary method of managing safety in most organizations. As a result, companies find their safety performance has plateaued and a high-performance safety culture has eluded them, producing exactly what the science predicts. This is why our advice is always to use primarily positive reinforcement to drive behavior change in safety. Negative reinforcement sometimes has a role in "jump-starting" behavior, but it should then quickly be replaced with positive reinforcement. Under positive reinforcement, employees always want, and try, to do more.

Recovery. A final cause of increases in behavior occurs when people who have been punished for unsafe behavior stop the behavior while the supervisor or manager is present, but resume it as soon as he/she leaves. This is the proverbial "when the cat's away, the mice will play" scenario.

For example, statistics show that many people leaving prison commit crimes on the very day they are released. Think about your behavior after receiving a speeding ticket. Most people resume speeding after only a few miles or the distance it takes to see the patrol car disappear from the rearview mirror.

All of this is to say that when an organization has improved its safety data, before you run out and celebrate, you had better understand the conditions that created the change. You might say that any improvement in safe behavior should be positively reinforced, and you would be right, but if you think what you are doing is positively reinforcing and it is not, you will never approach world-class safety levels.

PUNISHMENT, PENALTY, AND EXTINCTION: *THREE WAYS TO DECREASE BEHAVIOR*

Punishment. Punishment is any consequence that follows a behavior and decreases the frequency of that behavior in the future. Decreasing unsafe behavior is not the path to world-class safety.

There are always more ways to do something unsafely than safely. Therefore, without positive reinforcement for a safe alternative, the probability that punishing one unsafe act will lead to it being replaced by another unsafe act is relatively high. While it is possible to reduce unsafe acts with punishment, it usually creates only temporary benefits, which in the long run, aren't that beneficial.

Many undesirable side effects stem from a punishment-oriented management style. People lie, cheat, and falsify data to avoid punishment. They avoid responsibility and blame others. They escape through absenteeism and turnover. (We will discuss the dangers of punishment in greater detail in Chapter 7: Punishing People Who Make Mistakes.)

In the final analysis, punishment has little place in safety. If we increase behavior to 100 percent safe, what will be left to punish? That said, dealing with lying, cheating, and falsification of data is a no-brainer. In fact when these behaviors occur, the person should be fired. The reason for such a drastic action is that these behaviors are difficult to observe. If you don't know when behaviors are occurring, you can't apply consequences to them. Therefore, when employees are caught doing any of these things, it is better to let them find another place to work where, hopefully, they can correct these bad habits and escape the suspicion they would always be under in the present workplace.

Penalty. Penalizing employees for unsafe behavior is a poor way to teach them the correct or safe behavior. Suspension has been known to be a positive reinforcer for some employees. Demotion often creates a lingering resentment toward the company and the management. A system where one person's unsafe behavior will cause the loss of points toward a group celebration or loss of some tangible item often results in hard feelings toward the offender, causing teamwork to suffer. Not surprisingly, some people lie or cheat to avoid the penalty.

Extinction. Extinction is a strategy for eliminating bad habits by withdrawing reinforcement. In safety, extinction is seldom practical because it requires the user to have control of the reinforcer for the at-risk behavior. Many at-risk behaviors are maintained by natural consequences that cannot be eliminated easily (for example, working without PPE is more comfortable, easier, and faster). Furthermore, extinction takes too much time.

Extinction is important to understand in safety because of the negative effect it has on safe behavior. Many safe behaviors undergo extinction because of a lack of adequate reinforcement. In general it will take hundreds of reinforcers to get a safe habit to the level where it will continue indefinitely. If a new or improved safe behavior does not get sufficient reinforcement to become fluent, it will quickly stop.

In the next chapter we will highlight what makes consequences effective and discuss the role of antecedents in changing behavior.

---

[1] Wikipedia

[2] The term *behavior-based safety (BBS)* was originated by Dan Petersen, author of 17 safety books who, before his death in 2007, was considered the best-known safety professional in the U.S.

[3] Pierce, W. D., Epling, W. F. (1999) *Behavior Analysis and Learning*, 2nd edition, Prentice Hall Inc., Upper Saddle River, NJ.

CHAPTER 2

The Science of Behavior

Effective Consequences and the Role of Antecedents

The saddest aspect of life right now is that science gathers
knowledge faster than society gathers wisdom.
– Isaac Asimov

The first step to improved safety leadership is learning the effects
and side effects of each type of consequence. The second step is
learning how to use consequences for maximal impact. Again, the
science provides guidance.[1]

Characteristics of Effective Consequences

Understanding and changing behavior starts with a close look at
consequences. When assessing the consequences of frontline, at-
risk behavior, many leaders are puzzled. After all, the most obvious
consequence for engaging in at-risk behavior is getting hurt. Why
would anyone do something that leads to personal harm, even
death? Such cursory analyses lead to statements such as "She must
want to get hurt" or "He has a total disregard for safety." But some
critical scientific principles shed light on such behavior.

First, it is important to acknowledge that every behavior has
multiple consequences, usually a mixture of positive and negative

consequences (as defined by the performer). Second, the power or strength of each consequence is determined, not by how "big" or important it might seem, but rather by the timing and probability of that consequence. Thus, every behavior has a mixture of consequences; some positive, some negative, some immediate, some delayed, some high probability and some low probability. The *pattern* of consequences determines the performer's behavior. If the pattern of consequences favors the at-risk behavior, then the at-risk behavior will occur. If the pattern favors the safe behavior, then the safe behavior will occur. The ability to accurately analyze consequences leads to the ability to predict and change behavior. (Learn more at www.aubreydanielsblog.com, "Managing People for Maximum Performance.")

We categorize consequences using the following model to help our clients assess the impact that existing consequences have and then modify consequences to effectively drive important behavior change.

> *Type* of Consequence:
>
> > **Positive** to the performer
> >
> > **Negative** to the performer
>
> *When* the consequence follows the behavior:
>
> > **Immediately** while the behavior is happening or immediately after
> >
> > **Future** a few minutes, hours, days, or longer
>
> *Probability* of the consequence occurring:
>
> > **Certain** the consequence will happen close to or at 100 percent
> >
> > **Uncertain** the consequence only happens some of the time

POSITIVE VERSUS NEGATIVE

A positive consequence is one that encourages more of the same behavior and a negative consequence (not to be confused with a negative reinforcer) is one that discourages more of the same behavior.

Whether a consequence is positive or negative varies from one individual to the next. We are all different in terms of whether particular consequences will have a positive or negative impact on us. For example, some people love public praise and will work to get more of it. Other people are embarrassed by it and will work hard to avoid any recognition in front of others.

IMMEDIATE VERSUS FUTURE

Consequences that are immediate are much more powerful than those that are future. The further away a consequence occurs in time following a behavior, the weaker is its influence on behavior. This is one of the reasons that the threatened consequence of chronic back pain is not very effective at persuading young, strong employees to bend and lift properly or ask for help when lifting. Young people are relatively certain they will not hurt their backs the next time they lift something. Back injuries are often cumulative, so they may develop back problems after years of improper lifting, if they develop problems at all.

On the other hand, the immediate consequences for improper lifting (from the employee's point of view) include such immediate payoffs as saving time and effort, comfort, and looking strong and independent (if the alternative is asking for help).

CERTAIN VERSUS UNCERTAIN

Consequences that are certain are much more powerful than those that are uncertain. A good example of a certain consequence is burning your hand if you touch a hot stove. This will happen every time you touch a hot stove. Anyone with children knows it usually takes only one experience for a child to learn not to touch the stove again. In this case a painful injury proves to be a powerful consequence because it is certain and immediate.

A good example of an uncertain consequence is getting a speeding ticket. You could speed day after day for years and not get a speeding ticket. Getting pulled over and receiving a ticket is uncertain. Anyone who drives in most U.S. cities knows that speeding tickets aren't effective at getting everyone to drive the speed limit.

PICS AND NICS

Putting all the pieces together, we can analyze the power or strength of any consequence by determining whether it is positive or negative, immediate or future, certain or uncertain. As the following diagram shows, consequences that are both immediate and certain (regardless of whether they are positive or negative) are the most powerful.

Consequences that are positive, immediate but uncertain, are the next most powerful. Think of winning at a slot machine or catching a fish. Both of those are P, I, but U and yet they are very powerful consequences. Consequences that are negative, immediate, but uncertain are less powerful. Think of using intermittent punishment with your kids. It's less effective, right? Those consequences that are certain but future (whether positive or negative) are also less powerful. Finally, consequences that are both future and uncertain are the weakest of all.

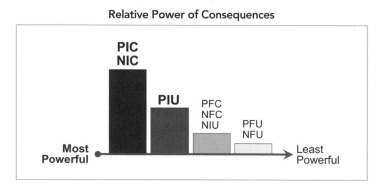

Relative Power of Consequences

16

PIC/NIC ANALYSIS®

Aubrey Daniels International (ADI) created the PIC/NIC Analysis as a simple way to do a functional analysis of any behavior. In safety, the analysis is particularly helpful for understanding the causes of at-risk behavior. It usually shows how at-risk behavior is being supported by unplanned consequences and how safe behavior is usually being inadvertently punished. The PIC/NIC Analysis is very helpful in illuminating the various sources of consequences and demonstrating their relative impact. It also informs how to arrange consequences that favor safe over at-risk behavior.

In many of the analyses we have done with clients, at-risk behaviors have been found to be reinforced by management and peers who have a history of reinforcing working quickly without enough attention paid to safety. We also frequently find many barriers to safe behavior that make it difficult or impossible for performers to do the right thing. The PIC/NIC Analysis will bring these to light. It is often instructive to do a series of PIC/NIC Analyses to illuminate the interconnectedness of antecedents and consequences in an organization.

Here is a common example. Frontline employees sometimes speed while operating equipment. A PIC/NIC Analysis shows the PICs that exist for the speeder.

PIC/NIC Analysis
Operator At-Risk Behavior: Speeding while operating equipment

Consequences	P/N	I/F	C/U
Get work done faster	P	I	C
Going fast (more fun)	P	I	C
Avoid being teased by peers for being too slow	P	I	C
Stay on schedule	P	I	C
Praise from boss for productivity	P	F	C
Get hurt	N	I	U

As you can see by this analysis, there are four PICs and one PFC for speeding. There is only one negative and it is uncertain, therefore weaker. Closer inspection shows that three of the positive consequences for unsafe behavior are related to better productivity. This tells us that there are probably many intentional and unintentional ways that working quickly (and therefore getting more work done—or perceiving that you are) is reinforced in this workplace.

Does that mean the supervisor is to blame? No. Below is a PIC/NIC Analysis on the supervisor's behavior. A common at-risk behavior that supervisors engage in is praising workers for high productivity (or meeting deadline) without checking to make sure the work was done without taking safety shortcuts.

PIC/NIC Analysis
Supervisor At-Risk Behavior:
Praising operators for productivity without attending to safety

Consequences	P/N	I/F	C/U
Feel like I am managing productivity	P	I	C
Getting product out	P	I	C
Easier to manage by results	P	I	C
Stay on schedule	P	F	C
Job done and no one was hurt	P	I	C
Increased incentive money	P	F	U
Praise from my boss	P	F	C
Avoid conflict about safety issues	P	I	U
Operator gets hurt	N	F	U

In this analysis, like the operator analysis, there are more positive consequences than negative for making safety communications secondary and the positives are more powerful than the one negative. The supervisor (like the operator) is choosing the at-risk behavior because the consequences support doing it.

The analysis can be done at all levels of the organization. Let's look at what might be going on at the higher levels where decisions are being made about organizational systems such as incentive systems, which are implicated in the supervisor analysis.

PIC/NIC Analysis
Executive At-Risk Behavior:
Weighing productivity much higher than safety in an incentive system

Consequences	P/N	I/F	C/U
Seeing good production report	P	I	C
Better long-term productivity	P	F	C
Increased profits	P	F	C
Company stays competitive	P	F	C
Shareholders are happy	P	F	C
May increase accidents	N	F	U

In each case, the performer (operator, supervisor, executive) is doing what makes sense given the consequences operating on their behavior. They are doing what appears to be the best for the company; after all, companies must produce to stay in business. As you can see, it is all too easy for leaders to have good intentions regarding safety, but to unintentionally have a negative impact.

Our purpose in this book is to help you see these interconnected contingencies as they are currently operating in your organization and provide suggestions on how to change them to improve safety.

(In Chapter 11 "Using Science to Understand At-Risk Behavior: PIC/NIC Analysis®", we will demonstrate another use for the PIC/NIC Analysis in accident/incident investigations.)

Characteristics Of Effective Positive Reinforcement

We have discussed characteristics that make all consequences more or less effective. Since positive reinforcement should be the primary consequence used in safety, this section will focus on how to use it most effectively with a brief review of the four characteristics that make a successful positive reinforcer.

PERSONAL

A positive reinforcer must be meaningful to the recipient. Of course, the way we know something is meaningful to a performer is by the response to the attempted reinforcer. If the behavior in question increased, the reinforcer was meaningful; if it did not increase, the reinforcer was not meaningful.

CONTINGENT

The word *contingent*, in a behavioral context, refers to the relationship between behavior and the delivery of a reinforcer. Technically, you do not "give reinforcers," people should earn them. It is helpful to ask, "What did the person do to *earn* this reinforcer?" When reinforcers are earned, they create feelings of competence and confidence. When they are not earned, they may eventually lead to feelings of dependence and entitlement. Reinforcement that is non-contingent (not earned) will not increase the target behaviors.

IMMEDIATE

Immediacy refers to the fact that when reinforcers are delayed, it is likely that the delay may cause you to inadvertently strengthen the wrong behavior because the behavior of interest is no longer occurring. This is why you may have heard about "catching people in the act" as the best time to reinforce. Positive reinforcers strengthen the behavior that is occurring when the person receives the reinforcer, not necessarily the one that you intended to reinforce. If someone is arguing when you tell him or her that s/he is doing a good job, you may strengthen arguing although you most likely would not intend to do so.

FREQUENCY

Frequency refers to the number of times a behavior is reinforced. In safety, we are concerned with developing habits that are so strong and stable to the extent that disruptions, pressure, and other external changes or distractions will not interfere with the safe execution of a process or procedure. Most safety professionals greatly underestimate the number of reinforcers required to develop fluent performance. By fluent performance, we mean "automatic non-hesitant responding." Once a behavior reaches fluency, very little reinforcement is required to keep the habit intact, but reaching fluency requires many more reinforcers than most people think. Various safety books advise that it takes 21 repetitions to develop a habit. Well, the research tells us that with such a few reinforcers, the behavior may last only a short time and will require considerably more time and effort (reinforcers) to maintain its integrity. The latest research[2] indicates that the number is actually in the hundreds.

When positive reinforcement is created following these rules, great things can be accomplished in all phases of corporate life. When they are ignored, not only will you be unlikely to get the results you want, but you may get more of the wrong results.

Antecedents

We mentioned that there are two ways to influence behavior. One is with consequences. This is certainly a most powerful influence and where the focus of behavior change generally should be. However, scientific knowledge is not complete without an understanding of the other influence: antecedents. Senior leaders focus much of their energies on antecedents in working with employees.

An antecedent is anything that comes before a behavior that contains information about the past behavior/consequence relationship or one that is desired. For example, the smell of some foods makes you hungry and others make you sick. Seeing a sign saying, "Eye protection required in this area" may result in one person putting safety glasses on and another one ignoring it. You can be sure that the differences in responding are the result of each person's different experiences in the past. When antecedents aren't consistently associated with a consequence, the response to the antecedents is inconsistent. This means that although safety signage, meetings, and lectures have a place in any safety system, they are in no way sufficient. It is important that we post a sign notifying motorists that the bridge is out rather than having them find out for themselves by running into the river. Of course we all know of people who have ignored such signs and run into the river. What would cause one person to heed the sign and another to ignore it?

Rule-Governed Behavior

There are two kinds of behavior: rule-governed and contingency-shaped. Contingency-shaped behavior is behavior that is conditioned through direct contact with a behavioral consequence. A child learns not to touch a light bulb after touching it and burning his fingers. Rule-governed behavior would be learned by a parent saying, "Don't touch that bulb because it will burn you." While many children seem to learn only by experiencing things for themselves, most adults learn that it is more efficient to follow rules since they are able to learn from the experience of others. In situations where following rules results in positive reinforcement, not only is the person more likely to follow the rules, but also more likely to follow the advice of the rule giver. Parents who are able to accurately predict consequences for their children will have children who will readily heed their advice. Parents who say, "Try it. You will like it," are more likely to have children who obey them if the child does indeed like it. This points to the fact that when employees do not follow safety rules, they have not had a history of reinforcement for following rules in other situations. In such cases the organization has inherited an obligation to teach them by making sure that following rules produces positive reinforcement.

A sign that says, "Speed limit, 55" is usually ignored because most people who go over the limit never experience any negative consequences, but do experience the reinforcers associated with higher speeds. The problem is that not only do these people ignore the speed limit but they may come to ignore other signs as well. A "No Parking, Towing Enforced" sign sounds serious but for some who violate it, they may well discover that it is not enforced and they frequently park in those spaces because the space is likely available and it is convenient to their destination. For this reason it is important to remove signs that are not enforced consistently. A sign that says, "Bridge Out Ahead" that remains posted for months after the bridge has been repaired may result in some people ignoring that sign and others as well. In the final analysis, rules are learned and maintained by consequences. Employees will come to either follow safety rules or ignore them depending on the consistency with which consequences (positive or negative) are experienced.

It is critical to understand how much reinforcement you as a leader get for setting up and managing antecedent conditions. Such things as strategic implementation planning, developing tactical plans, training, communication, defining values, conducting morning safety reviews, supply chain design, engineering analysis of materials and machinery, supplier/vendor orientations, even post-accident investigations are all antecedents for safe, efficient, and effective work. These are critical antecedent areas of focus that consume much of the time you devote to safety. They are important, very important, to get right—but without equal attention and effort to setting up and managing consequences around safety, these antecedent tasks will not ensure that you will have a safe workplace.

Motivating Operations (MO)

MO refers to any environmental change that has two effects: 1) it increases the momentary effectiveness of a reinforcer, and 2) it increases momentarily the behaviors that have produced that reinforcer in the past. The clearest examples are situations where you have been either deprived of a reinforcer or satiated by it.

The more you go without water or food, for example, the more reinforcing they become. After forgetting bottled water on a long hike, coming upon a stream leads to drinking from that stream, a low-probability behavior when bottled water is available or if the hiker is not thirsty. Lack of other water and heightened thirst make the stream water more reinforcing.

Marketing is filled with examples of MO's. When technology companies release a hot, new gadget in limited quantities, they are attempting to manipulate the reinforcing value of the item by making it difficult to get. Showing a popular celebrity wearing a certain brand of jeans increases the reinforcing value of those jeans.

Motivating Operations allow us a way to create reinforcers under conditions where the event was not normally reinforcing. A piece of ribbon may not function as a reinforcer until you are told that everyone that finishes a task in an allotted time will receive one. In safety, hearing a senior executive talk about the importance of safety and having supervisors start every meeting with a discussion of safety often enhances the reinforcing value of participating in the safety process. The degree to which these events are effective is directly related to the trust and credibility you as a leader have established with your followers. When you have demonstrated that you do what you say you will do (in other words that your actions match your words) then your antecedents are more likely to be effective.

Reinforcement History

Finally, it is important to be aware that the prior history of your employees (the habits, thoughts, and feelings of individuals) play a role in how people respond in safety. Individuals' past experiences (in your organization and others) will affect their willingness to engage in certain behaviors around safety. For example, the willingness to speak up, identify unsafe conditions, question rules that make no sense, and provide feedback to superiors may be very difficult for an individual based on that individual's history of doing so. Many people have developed persistent "values-based behavior" such as following orders, not questioning authority, and doing whatever it takes to get the job done. Leaders are sometimes surprised that despite repeated requests for more proactive safety behaviors, they come up short. It helps to understand the behaviors that people have learned and brought to work with them. This is not to suggest such behaviors are unchangeable, (reinforcement histories are constantly evolving), but understanding the behaviors will help you design more effective strategies for change. Finally, be aware of your own reinforcement history. Check your habits of decision making and patterns of behavior that may not be effective. An entire application of behavior analysis is dedicated to improving decision making in the analysis of risk. Learn about it. Analyze what you need and change past habits through effective consequences as well as antecedent training and practice.

Summary

The knowledge about the science of behavior analysis fills many books and we will not be able in this context to relate all such information. However, it is our opinion and experience that the leaders who know the most about the science create the highest safety performance. Our goal in writing this book is that you will want to learn more about the science as it enriches your life and all that you come in contact with at work, at home, and at leisure. To that end, we have included suggested readings in the back of the book.

---

[1] Portions of this section are reprinted from *Removing Obstacles to Safety* by Judy Agnew and Gail Snyder and with permission of Performance Management Publications (PMP).

[2] Schmidt, Richard A. (1991) Motor Learning and Performance: From Principles To Practice. *Human Kinetics.* Champaign, Ill. p.215.

PART TWO

Seven Safety Practices that Waste Time and Money

One credo of behavior analysis is "Focus on what you want, not what you don't want." Following this credo leads more often to the use of positive reinforcement strategies and avoids the detrimental effects of negative consequences. It is also most likely to lead to desired results. Those who focus on getting rid of undesired behavior often find the undesired behavior is replaced with more undesired behavior. It is better to pinpoint behaviors that lead to optimal safety results and use positive reinforcement to ensure they occur.

While we believe this is exactly the strategy required in safety, we are going to violate that approach in the next section of this book. We are going to discuss the things you should *not* do.

The pace and competitiveness of business today leaves no room for ineffective practices. The Lean movement is all about eliminating waste in manufacturing. But what about the waste in safety? Too few resources are available to do anything in safety that is not optimally effective. Yet, many organizations waste time and money on safety practices that not only do not lead to optimal results, but actually work against safety.

As always, with the science of behavior as our guide, we will show you how to reduce the waste in your safety system.

In the next section we review seven safety leadership practices from a behavioral perspective and explain the elements that make each practice less than effective. Next we discuss why companies use these practices, ending each chapter with suggestions on what to do instead.

No matter how far you have gone on a wrong road, turn back.
– Turkish proverb

Seven Safety Practices that Waste Time and Money

Practice #1: Focusing on Lagging Indicators

The Deepwater Horizon had seven consecutive years without
a single lost-time incident or major environmental event.
– Lou Colasuonno,
Spokesperson for Transocean,
as quoted in the *New York Times*

Managing safety is difficult. Many managers and executives would admit that they have a much greater sense of control over other parts of their business than they have over safety. Why would this be so? There are several reasons but it is partly a function of the metrics used to manage safety. Metrics significantly influence how things are managed in business. When those metrics are lagging indicators (focused on results), management tends to be reactive. That means that management is always dealing with how to fix problems rather than preventing them in the first place. Leading indicators allow more proactive management.

At the senior levels of most organizations safety is measured by lagging indicators such as Incident Rate, Lost Time Rate, Severity Rate, and DART (Days Away/Restricted or Transfer Rate). These are all after-the-fact measures of how many people were hurt and/or how badly they were hurt. They are not good measures of what employees do to prevent accidents on a day-to-day or week-to-week basis and therefore do not facilitate the proactive management of safety.

Lessons From Quality

Measuring safety via incident rate is akin to measuring quality by customer complaints. How many unhappy customers are there for every one that actually complains? For the ones who don't complain, what were the defects in their product or service? Was the one defect that was reported the only one? Incident numbers won't tell you.

Organizations that have consistent high quality "inspect it in." In other words, they monitor the process. They know when an employee is having a problem with incoming quality, with equipment, or with the process. This allows the problem to be contained and corrected immediately.

As with quality complaints, safety incident rates do not tell you how many near misses (or unreported accidents) there are for every reported incident. More importantly, they don't tell you how many at-risk behaviors occur for every incident. As the following diagram shows, the lagging indicators are only the tip of the iceberg; many at-risk behaviors occur and many unsafe conditions exist for every one incident.

The Probability Pyramid

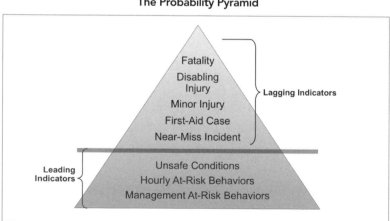

Most organizations have moved beyond using customer complaints as their primary measure of quality. Process measures and behavioral measures allow them to track quality throughout the manufacturing process and before their products get to the customer. These measures allow quality to be managed proactively to prevent problems and defects before they happen.

The typical proactive measures for safety include training classes completed, safety meetings held, and audits conducted. Unfortunately, these activities do not get to the critical questions: (1) Is the workplace as free of hazards as it can be? and (2) Are people doing things in a way that will prevent accidents or injury to themselves and others? Completing an audit does not ensure that unsafe conditions identified during the audit will get addressed. Attending a training session in proper lifting does not guarantee participants will consistently lift properly on the job.

Proactive Versus Reactive

Focusing on incident rate (and related measures) promotes a reactive approach to managing safety. That is, a great deal of the activity around safety is in *reaction* to incidents. Telltale signs of a reactive approach include periods of inactivity around safety followed by a flurry of activity when there is an incident or a near miss. Another sign is when frontline employees report that they only hear about safety in monthly safety meetings and/or after an incident. While many organizations try to be more proactive by talking about safety each day, that often translates into supervisors making general statements like "Work safely today" or "Be careful out there" at a start-up meeting. These antecedents are only a very small step toward a proactive approach.

For most managers (those above frontline supervision), safety is managed via the "no news is good news" approach. If there aren't any accidents or incidents, then the assumption is that everything is fine and employees are safe. While most managers would not like to admit this (because they truly care about the safety of their employees) a low incident rate allows them to focus on things that generate revenue and only casually attend to safety. If you doubt this, monitor the amount of time spent at your next review meeting discussing safety compared to the time spent discussing other business drivers. The average meeting, in which all business drivers are discussed, focuses heavily on issues such as production, schedules, sales, and quality. If safety is reviewed, the metric is usually some measure of accidents, and if that number is low or hasn't changed then there is a collective sigh of relief and everyone quickly moves on. Only if the number is high or has increased does the conversation concerning safety increase. As you will see, a reactive approach to managing safety is ineffective and does little to decrease the occurrence of incidents.

Frontline supervisors are in a different position because they typically spend more time observing hourly employees doing their jobs and are more likely to see the at-risk behaviors and unsafe conditions that can turn into accidents. Part of their job is to correct at-risk behavior. Again, this is a reactive approach. When an employee is seen without PPE, the supervisor provides corrective feedback. But what is being done to prevent this at-risk behavior in the first place? Such action on the part of the supervisor does not seem reactive, but it is, because the at-risk behavior has already occurred.

Incident Rate Is A Future, Uncertain Consequence

As we have discussed previously, the best way to improve performance of any kind is to provide positive, immediate, and certain consequences (PICs) for the desired performance. This is impossible if incident rate is your primary metric. It would be silly to have a celebration for every day that was accident free. Everyone knows that a company can go days, weeks, even months without an accident even if they are doing a lot of things wrong. Therefore, companies wait until a significant amount of time has gone by (for example, a million hours) because then it seems more likely that the lack of accidents is a result of working safely rather than luck. This means any celebrations around incident rate (days without a lost-time accident) are future and uncertain (not immediate and certain). Our point here is that the focus on incident rate as the primary metric drives the way safety is managed (with future and uncertain consequences). There is a better way.

Focus On the Negative

As noted earlier, when there are no incidents the natural consequences favor focusing on issues other than safety. When an incident does occur, management is moved to take action. But what form does that action take? Unfortunately, measures that focus on what is *wrong* favor the use of negative consequences. Once an incident occurs it is too late for positive reinforcement.

The measures we use often paint us into a corner. It is hard to get away from negative consequences when you only measure what goes wrong. The focus on the negative is not just around accidents. It is a natural consequence of supervision. Supervisors are not expected to respond to what employees are doing right (correctly) because correct performance is what they do most of the time. Therefore, supervisors look for errors in production, quality, and use of materials (waste). If you observe what most supervisors do daily around safety you will see that they look for and correct at-risk behavior. Managers and supervisors who have not been trained in BBS rarely look for and positively reinforce safe behaviors.

Is Safety Really Your Number-One Priority?

All of our clients state publicly that safety is their number-one priority, but they have a difficult time behaving that way. The messages that are stated publicly or on posters and those communicated daily are often different. Hourly workers frequently tell us that while the company says safety is number one, productivity is really number one. Furthermore, when we look at the amount of time supervisors and managers spend on safety versus other business drivers, safety does not appear to be number one. It is our belief that companies cannot live the stated value of safety as the top priority while using incident rate as the primary safety measure, because incidents occur too infrequently to enable management to make safety a priority focus. The annual incident rate for all industries in 2008 was 4.0 with a range of .1 to 14.0. This means that weeks and months could pass without an incident even in the most unsafe work environments. Productivity and quality win the priority battle because they have better, more sensitive, and frequent (often daily) measures that keep management's attention.

We will discuss the idea of *safe, efficient, quality production* later. That should be number one. Companies are not in business to be safe, because safety is not a product or service. Therefore, integrating safety into everything we do is the best way to really make safety number one.

Variation in Incident Rates Is a Management Trap

Let's assume that you just had an incident. It is thoroughly investigated and recommendations were made to correct the situation or behavior that produced the incident. Several months go by with no incidents. Did the actions that you took reduce the incident rate? Since months can go by without an incident, given your current rate, how long will you have to wait to know? Will you ever know for sure? Unfortunately, many managers are reinforced for such actions by the immediate lack of incidents. This creates superstitious learning. We have worked in plants where the rates were so low that a year or more may occur between incidents. We have seen managers fire employees involved in an incident and because months go by without another incident, they were inadvertently (superstitiously) reinforced for the action they took when in fact the lack of an accident was just the normal variation in the rate. As such, their action will have no impact on the future incident rate and probably a very negative impact on morale and future near-miss reporting. In most of the places we have worked, managers do not even know the mean and standard deviation of the number of days between accidents or incidents. These numbers will show you why you cannot manage by lagging measures. It is a statistical fact that if the yearly number of unsafe conditions and unsafe behaviors were held constant, an organization would experience a different number of incidents during the first half of a year and the last half (or from one year to the next). If you had fewer accidents in the last half of the year than the first, would that mean your safety process is now better than before? We don't think so but many organizations would use that data to prove that it was.

Why Companies Use Lagging Indicators

Companies focus on incident rate for several reasons. First, OSHA requires companies to report incident rate. It is OSHA's primary measure of organizational safety. The requirement to report to OSHA means the metric is readily available to use internally; so it saves time and money, both PICs.

Second, all other companies use incident rate as a primary measure. This allows comparisons across companies in the same or similar industries.

Finally, companies are often at a loss in determining better measures. Many safety professionals and managers understand the limitations of incident rate but struggle for meaningful alternatives.

What To Do Instead

 Moving to a proactive approach to safety requires a focus on what people in the organization are *doing*. Only when you see behavior can you actively manage a safety process. That being said, you must be careful not to confuse activity with a proactive safety process. As noted earlier, many organizations track safety meetings attended, safety training conducted, audits completed, communication of incidents, and so on. While these are not trivial activities and they should be tracked, they often produce minimal results because nothing of substance will change until behaviors change and antecedents are a very inefficient way to change behavior.

35

A combination of proactive, behavioral measures (folded into a matrix or scoreboard) should be the primary tool for tracking safety on a daily and weekly basis. By shifting the focus away from lagging indicators (which still need to be monitored), more effective safety practices will develop. Behavior-based safety provides proactive measures of safe and at-risk behaviors for the hourly population. This is a great first step, but it is equally important that supervisors, managers, and executives engage in measurable behaviors concerning safety that they can track on a daily or weekly basis. Tracking management behaviors and tying consequences to them is key to improving safety. Focusing primarily on frontline behavior will result in improvements that are short term. Leaders all the way to the top of the company must be fluent in behaviors that support the design and implementation of effective policies, processes, systems, and structures to ensure that safety is at the heart of all that is done. They must identify their own behaviors required to establish safety as core to all they do as well as the behaviors of those who report to them. Managers at each level must be able to promote *sustainable* behavior change, long before results are documented, whether at the front line or in the boardroom.

The last section of this book focuses on management behaviors and activities that can be easily turned into metrics (in some cases checklists are included in the chapters). Proactive leading indicators will give you a much truer sense of how safe your organization is, and they will allow more accurate and impactful use of consequences to drive improvement.

CHAPTER 4

Seven Safety Practices that Waste Time and Money

Practice #2: Injury Based Incentive Programs

*Rewards and incentives increase behavior
but they may not be the ones you want.
– Aubrey Daniels*

Peruse any safety magazine over the last decade and you will see many articles debating the value of incentives. Incentives have many proponents and have been very widely used in attempts to reduce incidents and accidents. (An Internet search of "safety incentive companies" results in 8,877,000 hits). Incentives, in fact, are a huge business. However, many opponents point to the problems incentives create, most notably the underreporting of accidents. In fact, OSHA is shining a renewed spotlight on safety incentive systems as part of their National Emphasis Program on Recordkeeping. At a recent conference, Jordan Barab, deputy assistant secretary of labor for OSHA said, "We will also be taking a close look at incentive programs that have the effect of discouraging workers from reporting injuries and illnesses. These include programs that discipline workers who are injured, or safety competitions that penalize individual workers or groups of workers when someone reports an injury or illness. *Let me underscore this point: OSHA will not tolerate programs that discourage workers from reporting injuries and illnesses.*"[1]

We'll talk about disciplining workers who are involved in accidents in a future chapter, but for now let's take a closer look at the role of incentives in safety.

What's Wrong With Incentives?

Nothing is wrong with incentives! It is the way they are used that causes problems.

Safety incentives often come in the form of some kind of merchandise, gift cards, or money. The merchandise comes in an ever-increasing range from low-cost items, (caps, logo jackets, and flashlights) to high-dollar items such as iPods, TVs, computers, and even automobiles.

The intent of a safety incentive is to increase safe working behavior resulting in decreased injury to the workforce. Who could argue with that outcome?

There are two general categories of incentive systems: those in which the payout is contingent on reducing or eliminating injuries and illnesses (we will refer to those as "injury based incentives") and those in which the payout is based on behaviors that proactively support working safely (we will call those "behavior-based incentives"). The controversy and concern revolves mostly around injury based incentives.

Some examples of injury based incentive programs include the following:

- Safety bingo (Individuals who do not have accidents participate in a bingo game spread over several weeks with an opportunity to win cash.)

- Group rewards offered for going a period of time without an accident (leather jackets at the end of the year if there are no recordables, for example)

- Bonuses based on incident rates (portion of salary as a bonus at end of year if there are no recordables)

- Prizes for individuals who work for a certain number of years without a lost-time injury (such as 5-, 10-, and 15-year, safe-driving awards).

In all of these incentive systems the criteria for earning the reward is not having an accident. Given that zero accidents is the ultimate goal of safety, the incentives sound reasonable on the surface. The problem is that the employees can get the incentives in three possible ways:

1. Employees *work safely* and thus earn the reward through desired safe behavior. In this case the incentives are operating in the intended fashion; they are motivating safe behavior and that safe behavior is preventing accidents.

2. Employees *engage in some or many at-risk behaviors* but are lucky in that none of the at-risk behaviors result in an accident. In this case the incentives are rewarding luck and possibly teaching employees that at-risk behaviors are okay: "It won't happen to me!"

3. Employees engage in at-risk behaviors and some of those at-risk behaviors result in accidents, but the *accidents are not reported* in order to avoid losing the incentive. In this case, incentives are motivating nonreporting of accidents.

This last point (number 3) is the one most often cited as the reason to avoid injury based incentives. (Learn more at www.aubreydanielsblog.com, "Do Behavioral Economists Really Understand the Behavior Part?") Motivating nonreporting is clearly extremely problematic and that alone is reason enough to discourage the use of such incentives. However, the second point is also troubling from a behavioral perspective. One of the big challenges in safety is convincing people that although their repeated at-risk behavior has not yet hurt them, it could.

As we talk to hourly employees about changing their at-risk habits we often hear, "I've done it this way for 20 years and I've never been hurt; why should I change now?" The fact that an at-risk behavior can be repeated hundreds or thousands of times without resulting in an accident is extremely unfortunate because it encourages the development of the bad habit. From a behavioral perspective, the consequence of getting hurt becomes more and more uncertain from the performer's perspective each time the at-risk behavior is engaged in and the performer does not get hurt. The more uncertain the consequence becomes, the less effective it is. Keep in mind the performer is also getting PICs for the at-risk behavior (it may be faster, easier, or more comfortable). The fact that a person can earn an incentive on top of these natural, positive consequences makes the situation even worse, as the incentive adds one more reinforcer that strengthens the at-risk behavior.

Obviously, the best scenario is number 1. But how can we know which of the three scenarios is playing out? In reality, probably all three are happening to some degree in most incentive systems. However, if you are not tracking safe and at-risk behavior then how can you know? If you are not tracking behavior, the chances are that you are inadvertently reinforcing risky behavior. Why would you want to risk underreporting and reinforcing luck when there is a better way?

A Simple Solution

We can make this very simple: injury based incentives are a bad idea. If you don't have such a system and are considering one, don't! If you have an injury based incentive system already, create a careful plan for eliminating it (see below). There are much more effective ways to motivate safe performance that don't encourage non-reporting and don't reward luck.

Why Do Companies Use Injury Based Incentives?

Incentive systems are always born from good intentions. Organizations that use them want to improve safety performance using positive strategies. Unfortunately, they often equate incentives with positive reinforcement. We hope that by this point you understand that the two are not the same. However, if managers don't understand what makes an effective reinforcement system (and injury based incentives are *not* effective reinforcement systems), an incentive is seductive. Their use is further justified by the fact that many companies use them (which lead to the assumption that they must be good). In addition, employees typically like them. It is important to separate whether employees like incentives from whether incentives actually achieve improvements in safety. In our experience, employees like getting the tangibles associated with incentives even if they know the incentives are not improving safety. When asked, most employees will admit that the incentives do not motivate safe behavior on a day-to-day basis but they don't want to lose them.

Another enticing reason is that there are hundreds of safety incentive companies who report millions of dollars saved by the implementation of an injury based incentive. How could this be so? Closer examination reveals that some of those gains are a function of suppressed reporting, and some of them come from a lucky streak. As we mentioned earlier, a rash of injuries is often followed by a streak of no injuries as a natural occurrence of the injury rate. In these cases, the money paid out in incentives is actually a cost rather than a saving because the results would have occurred without the incentives.

Another reason for the use of injury based incentives is the ease of implementing such systems. Since injuries are already tracked, it is just a matter of attaching incentives to them. In our society we are forever searching for the quick fix, but creating a high-performing safety culture is a complex endeavor. There is no pill; there is no pixie dust; there is no quick fix.

What To Do Instead

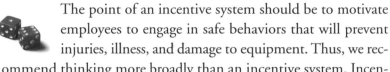

The point of an incentive system should be to motivate employees to engage in safe behaviors that will prevent injuries, illness, and damage to equipment. Thus, we recommend thinking more broadly than an incentive system. Incentives alone, however effective they may appear to be, are not the answer. Even behavior-based incentive systems are flawed when they focus exclusively on tangible rewards. Any time tangible rewards are offered (inside or outside of safety) there is a risk that people will lie, cheat, or steal their way to get the incentives. The larger the incentive, the more likely this is to occur. We are not saying incentives are bad, but they need to be managed very carefully in order to (a) get the desired behavior and (b) avoid undesired behavior (like pencil whipping or other kinds of data falsification). All these issues can be avoided by creating a reinforcement system, not an incentive system.

The reinforcement system may have some tangible rewards as part of it, but tangibles should only be a small part of the reinforcement system. A good BBS process has a solid, positive reinforcement system as a primary element. Listed below are some of the important pieces of a reinforcement system for safety improvement.

1. **Pinpoint safe behaviors and make sure they are directly linked to desired results**. For example, some behavior-based incentive systems focus on behaviors such as attending safety meetings. Does attendance at safety meetings link directly to reduction of accidents? Is it possible to have perfect attendance at meetings and still have incidents? Focusing on wearing Kevlar gloves and cutting away from the body, for example, are better pinpoints if cuts are common. When you focus on behaviors that directly impact accidents, your system is likely to be successful.

2. **Analyze your accident and near-miss data to identify (1) which safe behaviors might prevent your most common injuries, and (2) which safe behaviors might prevent less common but more severe injuries.** After determining the high-impact behaviors, validate your analysis by observing these behaviors in the work setting. It is quite possible that the incident records do not precisely pinpoint the critical behaviors.

3. **Engage people at all levels of the organization with pinpointed safety behaviors.** Creating a high-performance safety culture requires everyone in the organization to engage in behaviors that prevent incidents. While pinpointing behaviors at the hourly level is often more straightforward, it is critical to spend time identifying what staff, supervisors, managers, and executives can do, directly and indirectly, to prevent incidents and accidents. Those behaviors should be part of the reinforcement system so that everyone gets reinforced for working toward improved safety. This is often time-consuming because many people in these groups have not spent much time thinking about what they do that impacts safety. However, results show that the time spent in this activity is well worth it.

4. **Mix social and tangible reinforcers.** Saying "good job" or giving someone a thumbs-up every single time they engage in a safe behavior may produce satiation and as such reduce or eliminate the effectiveness of the reinforcer. Offering a gift certificate at the end of the month for those who have been consistently observed performing safely is also likely to be unsuccessful. It is the blend of social and small tangibles (with a heavy emphasis on social) that we have found to be most effective.

In particular, whenever possible, the social reinforcement should link the behaviors to their natural outcomes with statements such as "By lifting with your knees bent and back straight you are preventing potentially debilitating back injury" and "We are seeing fewer hand cuts since you started doing observations on wearing gloves." Interestingly, by using social and small tangibles, we find that natural reinforcers begin to take over. Employees begin to report a sense of pride when they engage in safe behavior, and they feel good when they help others work more safely. The ultimate goal in safety is to create a workplace where people engage in safe behavior because it is the right thing to do. The feeling that you are truly and tangibly helping yourself and others stay safe on the job can be a powerful reinforcer when we help people tap into it. The recipe to get there includes building in more immediate and certain social and tangible reinforcers.

In summary, if you link your reinforcement to behavioral data that shows what people are doing to prevent accidents and incidents, you will get more of these behaviors. When preventative behaviors increase, incident rate will take care of itself.

How To Eliminate An Injury Based Incentive

If you currently have an injury based incentive system and you would like to eliminate it, you have a difficult job on your hands. No matter what you do, many people will view it as a take-away and will be unhappy. The best approach is to begin a behavior-based process that focuses largely on social reinforcement and run both systems simultaneously for a short time. Once everyone is comfortable with the behavior-based process and sees the value in it, ask them if they like the idea of taking the injury based incentive money and putting it toward the behavior-based process. In most cases people will see that a behavior-based approach is more likely to improve safety, and they will be willing to make the change. They also see that they have more control over the rewards or celebrations when results are based on their own daily behavior.

Ensure you are clear that there will be a shift in how the money is distributed. Rather than large, infrequent rewards (typical of injury based incentives), the money should be used for small, more frequent celebrations. People will miss the large rewards, but will come to appreciate the logic in the change.

As senior leaders you too operate under incentives based on scorecards, one component of which is usually safety. Look carefully at your system of incentives. Ethics investigations following large incidents have uncovered stunning stories regarding how those in charge of safe outcomes worked under systems in which they were rewarded despite not having created a safe workplace. Do your incentive systems truly target the behaviors that you must do to keep everyone safe? Make sure the metrics upon which incentives are based, and which ultimately benefit you and others at the top, focus on aligned and meaningful safety behavior at all levels, not just results.

---

[1]Barab, Jordan. (August 18, 2009) Speech given at the United Steelworkers Health, Safety and Environment Conference. Houston, Texas.

CHAPTER 5

Seven Safety Practices that Waste Time and Money

Practice #3:
Awareness Training

What we think, or what we know, or what we believe is, in the end,
of little consequence. The only consequence is what we do.
– John Ruskin

Training is a necessary but not sufficient component of all safety systems. Obviously employees need to be trained in the hazards of the job and in safety procedures. However, is the goal of the training to make people aware or to change their behavior on the job?

While the stated goal of some safety training may be simply to inform (make people aware), all safety training has behavior change as its goal. What is the point of showing a video on how to avoid slips, trips, and falls if not to have people change their behavior? The goal of training is never for participants to go back to doing what they have always done.

If the goal of training is behavior change, then by now you will have seen the flaw. Training, as a sophisticated form of telling, is an antecedent. Such antecedents rarely lead to long-term behavior change. Predictably, the most common outcome of safety training is a temporary change in behavior followed by a gradual return to old habits. The problem is not that people don't care or don't want to change. The problem is that awareness training is only an antecedent and does little to affect the consequences in the workplace where the behavior occurs. Thus, if those consequences favor doing at-risk behavior, seeing a video on the risks of that behavior will have short-term impact at best.

More Is Not Better

When training fails to produce long-term results, a common organizational response is to do more training. This will produce only bursts of desired behavior followed by the inevitable return to old habits. The answer lies not in antecedents but in consequences.

Let us be clear. We are not suggesting abandoning safety training. We are recommending reserving training for those circumstances for which it is the correct solution, and improving the effectiveness of training so that it targets skill development and fluency, rather than awareness.

Why Do Companies Do It?

Training is the most common solution offered for most safety problems. While some safety issues require more effective training, it is often not the solution. However, training does give concerned parties a sense of "doing something" about a problem. Furthermore, management behavior of approving training as a solution may be reinforced when the training results in short-term changes in behavior. Unfortunately, long-term impact is often not assessed.

Training is only a small part of the solution to a complex behavioral problem. The desired impact of safety training is to get a behavior started that can then be positively reinforced in the workplace. Those who understand behavior can see that the analysis and adjustment of the consequences for problem behaviors is where long-term solutions lie.

What To Do Instead

 Before you resort to training as a solution to a safety problem consider the following:

1. Establish whether the problem is truly a training problem. The TRAC model below was developed by a colleague, Dr. Cloyd Hyten, and is a modification of Tom Gilbert's[1] Behavior Engineering Model. It shows that training is only one of four major causes of performance issues.

The Big 4: TRAC

Training	Resources	Antecedents	Consequences
Train efficiently	Proper tools/ equipment available	Set attainable goals	Emphasize Positive Reinforcement:
Share best practices	Insure safe & quality work procedures	Clarify expectations	Add PICs for desired behaviors
Use job aids to shorten training	Insure adequate staffing	Add clear prompts	Reduce NICs
Use fluency to make training more effective		Use proper feedback	Align pay, promotion, evaluation systems to support performance
Insure follow-up		Reduce distractions	
Evaluate by impact on performance		Clarify metrics	Eliminate conflicts between incentives
		Reduce stimulus confusion	

The first step is to determine whether the problem is a result of inadequate training. For that we recommend conducting Robert Mager and Peter Pipe's "Can't Do/Won't Do" Test.[2]

The "Can't Do/Won't Do" test is a simple one. If the performers couldn't do a behavior or task if their lives depended on it, then it may be a training problem. But as the model above suggests, it may also be a problem of resources or a problem of poor antecedents. In "can't do" cases, giving performers an incentive, paying them more money, or threatening them with loss of their jobs will not solve the problem. The performer simply cannot do it.

If the performer(s) can do the behaviors if their life (lives) depended on it but don't, it is a motivational problem—a problem of consequences. If employees have done the behavior in the past but no longer do it, it is still a problem of consequences. The problem could be lack of a reinforcer or in some cases satiation on past reinforcers.

In our experience, most safety problems are a function of resources and consequences. Only occasionally is training the problem.

2. If you find that the problem is a training problem, aim for fluency as the training outcome, not awareness. *Fluency* means the performer can complete the task accurately and without hesitation. Awareness training will never produce fluency. Too often, training involves a more seasoned employee showing a new employee "the ropes." This is usually awareness training in that no action is required on the part of the learner. They just listen. If there are certain behaviors or procedures employees need to be able to do after training, then built-in deliberate practice is required.

Data-based, proven, instructional design technology can train employees to the point of fluency in less time than traditional training methods. In a training program developed for customer service representatives, we included over 800 customer problems for trainees to solve. The previous training included less than 50 customer problems. The results were that training time was cut by 34 percent from nine weeks to less than six weeks and trainees were performing at the level of seven-year veterans after week two of the training. Two supervisors in a distribution center, who were trying to improve the efficiency and quality of loading trucks, checked designated behaviors over 500 times in a two-week period. The feedback and reinforcement associated with the checking resulted in record performances. Such fluency training applied to safety would yield equally impressive results. (For more reading on fluency, see articles by Carl Binder and Kent Johnson & Joe Layng in the suggested reading list in the Appendix.)

3. Create a follow-up plan that includes evaluation of behavior change after the training. If performance has deteriorated, you either did not train to fluency, do not have an effective reinforcer, or have reinforced too infrequently. Revisit these elements and continue to do so until the desired level of safety is maintained.

4. If you determine that the problem is not a training problem, look to resources, other antecedents, or consequences as the source of the problem. A PIC/NIC Analysis of the current and desired behavior should highlight the source of the problem and point to potential changes that will remedy the problem.

---

[1] Gilbert, T. (1978) *Human Competence: Engineering Worthy Performance.* New York: Mc-Graw-Hill

[2] Mager, R.F. & Pipe, P. (1999) *Analyzing Performance Problems or You really oughta wanna.* Belmont, CA: Fearon-Pitman Publishers, Inc.

Seven Safety Practices that Waste Time and Money

Practice #4: Safety Signage

> Publicly or legally reminding people of their responsibilities may
> have some effect in getting them or others to behave differently
> (though never for a long time).
> – Sidney Dekker

We will review three types of safety signs: compliance signs, information signs, and inspirational signs.

Compliance Signs

Compliance signs communicate the conditions under which specific behaviors should occur. Here are some examples:

Danger, explosive vapor

Wear breathing apparatus in this area

Authorized personnel only

Eye protection required

Hearing protection required

Foot protection required

Like safety training, compliance signs are a necessary but not sufficient part of any safety system. They are necessary because they clearly communicate important information (not to mention that OSHA requires them) but they are not sufficient in that posting such signs does not guarantee that employees will engage in the target behaviors.

We are not suggesting eliminating compliance signs since they serve an important function. We are suggesting being realistic about their effectiveness. Signs alone won't change behavior; if they did people would never speed on freeways, smoke cigarettes, or take more than two pieces of carry-on luggage on airplanes.

Why don't people respond to signs? Some think it is because they didn't see them (or possibly they can't read). Usually the second attempt at using a sign to change behavior is to post a larger sign and/or to place the sign in a more strategic position. The assumption behind this activity is that people do what they are told; therefore, all we have to do is make sure that they have been told.

The problem with signs is that everyone doesn't respond appropriately to the information on the sign. If you think about it, compliance signs limit personal freedom in some way. They ask you, or require you, to drive slower, obtain permission, limit access, or in some other way they impede your activity or burden you with extra work (as in requiring a permit)–all NICs.

In the final analysis, the reason people don't always respond to compliance signs is because the consequences don't favor it. Many people will tell you that they didn't see the sign, and that may be true. Ask any employee and he/she will tell you that after a few days or weeks, a new sign blends into the wall and they don't notice it anymore. But the reason they didn't see the sign is usually because there was no reinforcement for seeing the sign. If we had a sign that read, "Free ice cream in the break room," how many people would respond? If you posted a sign that said, "Everyone who is standing in this area at 3 o'clock can go home," you would probably create a stampede.

Therefore it doesn't appear to be the sign that is the problem. It is the consequences.

Information Signs

Information signs are intended to communicate something about safety that is not compliance related. Examples include signs announcing new safety training, the availability of updated safety gear, or the kickoff of a new BBS process. In these circumstances, signs are intended simply to inform or to generate one-time behavior (such as signing up for the new training or securing the new safety gear). As with all antecedents, signs may function effectively to generate one-time behaviors. In the case of a sign that kicks off a new safety program such as BBS, signs can be a helpful component of a broader communication plan. Attractive new signs often catch people's attention and serve as great prompts for discussion and a call to action. Keep in mind that these signs, like all others, will have short-term impact, so plan to capitalize on the temporary effect and then move on. Leaving any signs up that are no longer relevant or "old news" only encourages the ignoring of signs.

Inspirational Signs

The third type of safety signage we call "inspirational." Rather than focusing on specific behaviors, these signs are an attempt to encourage safety in general. Here are some common examples:

- Think Safety
- Chart a Course for Safety
- Safety Starts with Attitude
- Walk the Walk/Talk the Talk, Safety First
- Make Safety a Daily Mission
- Safety Depends on You
- Start your Shift with Safety In-Gear

We assume the goal of these signs is to get people to "think about safety." For this purpose, the signs are marginally effective, at best. It is likely that a person looking at one of these signs for the first time (and a few times after that) will think about safety. The question is what specifically will they think about and, more importantly, will that result in safer behavior? Even if some change in behavior occurs, the science of human behavior tells us it will be a short-term change. Inspirational signs do little to improve safety. Save your money.

Why Do Companies Do It?

As with many of the practices reviewed in this book, organizations use safety signage with good intentions. Here are some of the reasons signs seem like a good idea:

- There is a misplaced belief in society in general that signs change behavior. Putting up a sign is often the first solution offered for behavior problems. Most companies use signs which lead to the assumption that the signs must be effective.

- Management often views signs as a way to visually and concretely demonstrate their commitment to safety. Our advice: demonstrate your commitment through your behavior, not only through signs.

What To Do Instead

As noted, compliance signs are important. We advise you to use only compliance signs that direct specific behavior ("Hearing protection required in this area") in order to maximize effectiveness. Use informational signs when appropriate but be sure to take them down when they are no longer relevant. We recommend that you stop using inspirational signs. Without the clutter of signs that have no meaningful information, employees may be less likely to ignore important signage. Understand that, at best, signs will have a temporary effect on behavior. Plan positive consequences to reinforce the behaviors identified on the signs.

Rather than inspirational signs, post behavioral feedback graphs that show employees their safety performance. Graphic displays of data on safe performance are much more likely to have a positive impact on safe behavior.

If you are looking for a way to demonstrate your commitment to safety, do so through your behavior.

- Talk about safety daily; ask about safety as you interact with others.

- Reinforce safe behaviors each and every day and you will be demonstrating your commitment to safety and having a positive impact on safe behavior. (See Part 3 of this book for an in-depth discussion of leadership behaviors that demonstrate commitment.)

- Finally, remember that in safety, as in everything else, lasting improvement happens through consequences, not antecedents. (Learn more at www.aubreydaniels-blog.com, "Talk Does Not Cook Rice.")

Seven Safety Practices that Waste Time and Money

Practice #5: Punishing People Who Make Mistakes

How very little can be done under the spirit of fear.
– Florence Nightingale

Consider the following:

1. An operator fails to follow procedure and check the pressure on a tank every two hours. Unnoticed, the pressure increases, causing damage to the tank and nearly resulting in the release of a toxic chemical. Other operators had also failed to follow the proper checking procedure. Supervisors were aware of this noncompliance but did nothing. The operator involved in the incident was given a 30-day suspension.

2. After the breakdown of a critical piece of equipment, in an effort to get the job done more quickly, a mechanic fails to fully lock-out and tag-out the power source. Although tremendous pressure was put on the mechanic (by both peers and management) to get the equipment running again, nothing was said about

safety at the time. The mechanic was formally repri-
manded with a written warning.

3. A worker was struck and killed by an earthmover on a
   construction site. Even after the investigation it was a
   mystery why the worker was in harm's way. The project
   manager was fired for "not creating a culture of safety."

To Punish Or Not To Punish

Given the seriousness of the above scenarios, many would argue
that in every case punishment was indicated. Punishment, or *discipline* (*see next page*) as it is usually called, is considered an important part of most safety programs, but organizations often struggle
with identifying the circumstances under which it should be used.
Some companies use it sparingly. Others have "zero-tolerance"
policies which lead to more frequent use of punishment.

Many issues should be considered when deciding whether
punishment is the right response to accidents, incidents, near
misses, and at-risk behavior. Many believe that the effects of punishment are straightforward. They are not. Our purpose here is to
explore punishment from a scientific perspective to enable you to
make better decisions about how to respond to safety incidents. A
good place to start is by answering the following important questions:

- What is the desired outcome and will punishment really
  lead to that outcome?

- Has punishment resulted in the desired outcome in the
  past?

- What side effects might punishment have, and are the
  benefits worth the cost?

- Can you accomplish the same outcome without punishment?

DISCIPLINE OR PUNISHMENT?

The terms *discipline* and *punishment* are often used interchangeably and in fact we will do so in this chapter, even though their scientific definitions are different. Given the difference in meaning between common usage and scientific usage, however, it is instructive to define them.

The scientific definition of *discipline* refers to the process of creating a disciple, who is a person able to self-manage his/her behavior in a prescribed way even when no one is watching. To discipline, therefore, means to instruct a person to follow a particular code of conduct or order. It is easy to see why the term was adopted in safety. This scientific definition is exactly the goal when someone makes a mistake in safety. Unfortunately, this is rarely the outcome of the "discipline" practices used in most organizations. In most cases discipline is used in response to an unsafe behavior and is intended to decrease the probability of that behavior being repeated.

Punishment is defined scientifically as any consequence that follows a behavior and reduces its frequency. In common usage *punishment* refers to acts that are *intended* to reduce behavior. The important distinction is the *a priori* vs. *post facto* evaluation of whether punishment has occurred. In behavior analysis punishment can only be determined after a decrease in behavior (post facto). In common usage, punishment is determined before the application (a priori). It is assumed that certain events (being written-up, chewed-out, and suspended without pay) will be punishing to everybody. The scientific fact is that these actions may not be punishing to some people and, in fact, may be positively reinforcing to others.

To those not familiar with the science of behavior these differences may seem trivial; however, they are often critical to dealing effectively with behaviors that may cause accident and injury. In this chapter we use the common definitions of the terms *punishment* and *discipline* since those definitions most closely reflect most readers' experience with these terms. However, when we later discuss problem analysis and application, we will use the scientific definitions.

Accidents Have Multiple Causes

A common response to an accident is to find out what caused it and, if it was human behavior, to punish that behavior to ensure it doesn't happen again. This approach has several problems, one of which is that accidents usually have many causes. The root cause process used by one of our clients includes over 200 different possible contributors to be considered when investigating an incident, for example.

With the many ways that things can go wrong, it is unlikely that the cause of an accident can be boiled down to one person's behavior. Most accidents represent the coming together of multiple events or circumstances, any one of which alone would not result in an incident. Unfortunately, in too many cases, despite the multifaceted nature of the event, investigations end in discipline of the frontline employee(s) at the point of the accident. As we will discuss below, this simplistic response rarely has the intended impact.

Furthermore, when other employees see or suspect that the accident has multiple root causes and that some of the blame lies in management-controlled circumstances (hazardous conditions, poor design of equipment, motivation systems, and so on) resentment builds and moves the organization further from a high-performance safety culture.

Effective Punishment

The definition of a punisher is anything that follows behavior that reduces the probability of that behavior in the future. Given this definition, punishment appears to be the right consequence after an incident. If an at-risk behavior causes an accident (or is one of the causes), it is reasonable to want to reduce the probability of that behavior. Yet, using punishment effectively is anything but simple.

Several variables impact the effectiveness of punishment and thus make it difficult to use. First, as with reinforcers, things that are punishing vary widely across people and circumstances, so standardized disciplinary processes are not likely to work universally.

Second, a delay between the undesired behavior and the punisher reduces the punishment's effectiveness. The longer the delay, the weaker the effect. This is one of the reasons that the natural negative consequence from an injury is a better punisher than a reprimand that comes days later. Third, research shows that occasionally punishing undesired behavior is much less effective than punishing many or all instances of the undesired behavior. Thus, when at-risk behavior is occurring often, punishment delivered only when the at-risk behavior causes an incident is too infrequent. As you can see, even when punishment is warranted, it is very difficult to use effectively in a work setting.

Side Effects Of Punishment

In addition to the difficulties of executing punishment effectively, one must deal with its predictable negative side effects. To use punishment without considering the side effects is like using thalidomide to reduce morning sickness. It will work, but at what cost?

The side effects of occasional and appropriate use of punishment are mitigated when embedded in a culture of positive reinforcement. Unfortunately, most organizations use too little positive reinforcement to offset the negative side effects of punishment. Listed below are the common side effects of the disproportional use of punishment:

- Lower morale
- Lower productivity
- Decreased teamwork
- Decreased volunteerism
- Increased turnover
- Lower trust
- Desire to retaliate
- Suppressed reporting of incidents, accidents, and near misses

While all these side effects should be of concern, the last one is of critical importance to safety. When people are fearful that reporting an incident, near miss, or even an at-risk behavior will result in punishment, they will not report. Without accurate, honest, and frequent dialogue between hourly workers and management, organizations cannot approach a sustainable, high-performance safety culture.

As consultants we have often witnessed the surprise expressed by management when an incident occurs. The surprise comes from not being in a position to see the circumstances that set up the incident (the coming together of root causes). In many cases the hourly workers are less surprised because they saw the precursors: hazardous conditions that didn't get reported (or got reported but not fixed); employees working beyond the point of fatigue; at-risk behavior patterns in themselves and their peers; antecedents and consequences that encouraged risk-taking, et cetera. Management is not in the work environment enough to observe or experience all of these potential root causes. A proactive safety culture requires hourly workers' willingness to discuss the obstacles to safety that they see and experience, but most importantly, a willingness of employees to work together with management to address such obstacles. Punishment suppresses that willingness.

Sidney Dekker, in his book *Just Culture: Balancing Safety and Accountability*, discusses how punishing those involved in accidents and incidents does more damage than good precisely because it leads to fear of reporting and discussing near misses and at-risk behaviors. He describes many situations where valuable information that could have prevented accidents was not brought forward out of fear of punishment. We'll discuss this more later. For now, let's look at why organizations use punishment or feel compelled to use punishment around safety.

Why Companies Do It

In our years in the safety business we have seen a variety of responses to accidents and incidents and a variety of reasons given for those responses. Some of the reasons for using punishment are described below. In the vast majority of cases the response an organization selects has, as its primary goal, ensuring the accident or incident does not happen again. As we shall see, if prevention is the desired outcome, punishment is rarely the right strategy.

REASON 1:
PUNISHMENT PROVIDES A SENSE OF "DOING SOMETHING."

A client in the early stages of implementing BBS experienced a rash of incidents. The executive vice president of operations, who was very supportive of BBS and understood the dangers of punishment, finally said, "But I have to do *something* about these incidents!" Our response was, "You *are* doing something." Using a BBS process to correct the conditions and behaviors that lead to incidents is *doing something*, but it takes time. His response, that it simply didn't feel like he was doing anything, is not unusual. We have heard similar sentiments from other clients. It is unfortunate that a single, decisive, negative action in reaction to an incident is considered to be "doing something" but a plan focused on ensuring the incident will not happen again is not. In reality the plan to strengthen safe work habits will have better and more lasting impact than any disciplinary action ever would. It just doesn't feel like it at the time.

REASON 2:
USING PUNISHMENT TO ENSURE THE PERFORMER DOESN'T DO IT AGAIN

One of the reasons given for using punishment is to make sure the employees directly involved in the incident do not do it again. We suggest that in the case of injury accidents, the injury itself is usually the best punisher of behavior. As anyone who has suffered a reasonably serious injury because of his/her own behavior can attest, you are very unlikely to do that behavior again. If the pain and suffering is immediate and serious enough, it will function very well to decrease the behavior, and reprimands and suspensions are unnecessary. If the incident does not result in injury, punishment may reduce the performer's tendency to do the behavior again, but is likely to create one or more of the undesirable side effects previously mentioned.

REASON 3:
USING PUNISHMENT TO SET AN EXAMPLE

Those using punishment often understand that any punishment administered to an injured worker is probably unnecessary, but believe that punishment is important to "send a message" to other employees. It does not seem just to punish someone in order to send a message to others. However, many people believe that not punishing would be to condone the behavior that precipitated the accident. In cases where the performer engaged in blatant and willful violation of policies (in other words, where it is clearly the intent to cause harm) termination is indicated. We will argue later, however, that intention is difficult to determine.

Regardless of whether it is deserved, using punishment to set an example is an attempt to manage the behavior of others through negative reinforcement. The message is, "If you want to avoid punishment, don't do this behavior." We have discussed the limitations of negative reinforcement earlier in this book. Dekker sums up the problem well when he states, "The idea that a charged or convicted practitioner will serve as an example to scare others into behaving more prudently is probably misguided: instead, practitioners will become more careful *only in not disclosing what they have done.*"[1]

REASON 4:
USING PUNISHMENT BECAUSE NOTHING ELSE WORKED

Organizations often believe punishment is required because other attempts to change at-risk behavior have failed. This is not surprising given that most attempts to change at-risk behaviors primarily involve antecedents. This is evidenced by the comments often heard after an incident: "We told them over and over that it is not okay to do this." "The policy is clearly stated in the hand-book." "There are signs posted in the area." "We just talked about this in last month's safety meeting." Signs, policies, training, meet-ings, and discussions are antecedents. Those who do not under-stand that antecedents will not change behavior permanently are often led to believe that punishment is the only way to "get peo-ple's attention." It's not.

REASON 5:
USING PUNISHMENT TO DEMONSTRATE ACTION TO EXTERNAL PARTIES

Sometimes punishment is not directed toward behavior change at all, but is simply a way to demonstrate that action has been taken. Serious incidents usually result in a call to action—"What are you going to do about it?" Punishment is sometimes used to show sen-ior management, OSHA, other employees, or family members of injured parties that action has been taken. Punishment looks like a good response. It rarely is. Most concerned parties would be sat-isfied with a plan detailing how similar incidents will be prevented in the future.

REASON 6:
USING PUNISHMENT WHEN OTHERS CALL FOR "SOMEONE TO PAY"

Sometimes the call for action is more specific: there is a call for those responsible to "pay for their actions." This is most common when injury or damage has occurred as a result of the incident (someone was hurt or sickened, property or the environment was damaged). Payment aimed at repairing damage may be appropri-ate. However, if the call for "someone to pay" is aimed at simply punishing the responsible party, what purpose does that serve? If it is to make sure that particular individuals won't do it again, it may or may not work, as we have discussed. If it is to make vic-tims feel better, it is probably not worth the cost. Most impor-tantly, punishment will not ensure the incident won't happen again and isn't that the most important outcome?

Misplaced Accountability

The call to hold someone accountable for an incident is ubiquitous. It is heard after airline accidents, medical errors, environmental disasters, industrial accidents, and even situations such as a country's financial crisis. Accountability is important. As Dekker points out, "Calls for accountability themselves are, in essence, about trust. They are about people, regulators, the public, employees, trusting that you will take problems inside your organization seriously. That you will do something about them, and hold the people responsible for those problems to account."[2]

The question is, Does accountability have to involve punishment?

Virginia Sharpe, in her studies of medical harm, has made an important discrimination between what she calls "forward-looking accountability" and "backward-looking accountability."[3] Backward-looking accountability is about finding blame, finding the individual who made the mistake and delivering punishment. As we have noted earlier, such action may feel good, and may ensure that an individual performer won't make that mistake again, but delivering punishment won't do much else. It won't guarantee that others won't make the mistake and it won't fix any organizational problems that contributed to the incident.

Forward-looking accountability acknowledges the mistake and any harm it caused but, more importantly, it identifies changes that need to be made, and assigns responsibility for making those changes. The accountability is focused around making changes—building safe habits and a safe physical environment—that will prevent a recurrence, not on punishing those who made the mistake.

As noted earlier, investigations usually have as a primary goal ensuring an incident doesn't happen again. However most of the effort is expended on backward-looking accountability. After the investigation is done, the appropriate parties have been disciplined, and any damage repaired, there is often spotty follow-through on the action items identified to prevent a reoccurrence. Concrete action items such as repairing a piece of equipment have a high probability of completion. It is the less-tangible action items such as changing supervisory practices, modifying processes for ensuring better engineering designs, and encouraging peer feedback around behaviors that lead to incidents, for which accountability often falls apart. This is precisely the accountability we believe should be the focus. Holding the appropriate parties accountable for fixing the behavioral conditions that lead to the incident should be of primary concern. Yet, such action is often dropped or done poorly. One act of punishment is easy; ongoing follow-up is not.

Lost Lessons

Organizations are always changing. Technology changes, the equipment changes, the people and their skill level changes, and the pressures and priorities change. This results in ever-changing hazards. A world-class safety process is one where people are responding to those changes and asking (out loud) how accidents might occur and how to prevent them. It is clear that extremely valuable lessons can be learned from near misses, incidents, accidents and at-risk behavior. Yet most organizations struggle to get workers to report near misses and worry about non-reporting of accidents.

Make no mistake: it is the threat of punishment that causes these problems. When people are afraid to discuss mistakes because mistakes too often lead to punishment, then they will not come forward and the lessons will be lost.

Does Lack Of Punishment Condone At-Risk Behavior?

As we have said, sometimes punishment is the right thing to do. Blatant, willful violations of safety policies and procedures are probably best handled with punishment. To not punish under such circumstances may indeed condone at-risk behavior. But what about the majority of circumstances that are not so clear-cut, where blame is not so easily assigned? If incidents are resolved without punishment, does that condone any at-risk behavior that may have been involved? We think not. We believe it is possible to clearly communicate that certain at-risk behaviors are unacceptable without the use of punishment. Or, more importantly, we believe it is possible to clearly communicate the criticality of certain *safe behaviors* without the use of punishment.

Priorities are communicated by what management does. When management works to change the antecedents, consequences, and conditions that lead performers to engage in at-risk behavior, a priority is established. When management works with the hourly population to establish effective safety improvements, it sends the message that safety is the priority. Relentless action toward preventing at-risk behavior sends a clear, consistent, and fair message: no punishment required.

What To Do Instead

ESTABLISH A PROCESS FOR DRAWING THE LINE BETWEEN *HONEST MISTAKE* AND *WILLFUL VIOLATION*

Once you are convinced of the limitations of punishment you still must decide when it is the right thing to do and when it is not. If punishment is to be reserved for blatant and willful violations, how will you decide what is blatant and willful and what is an honest mistake? This will always be a judgment call. In Dekker's words: "To think that there comes a clear, uncontested point at which everybody says, 'Yes, now the line has been crossed; this is negligence,' is probably an illusion."[4] Dekker suggests that *who* draws the line is the critical decision and recommends including "practitioner peer expertise" (peers who do the same work). This is important because, as noted earlier, management and safety personnel (those usually involved in investigations) cannot fully understand the behavioral and physical conditions under which the work is done. A peer can.

In addition, we suggest including an expert in behavior analysis.

It is far too easy and too common to conclude that if a person was trained in the safe behavior but did something at-risk instead, that it was willful and blatant. In our experience, when analyzed from a behavioral perspective, many cases of seemingly blatant safety violations can be shown to have systemic, organizational contributors to the behavior. These are often subtle things like pressure around productivity, fatigue due to overtime work, strong negative consequences for the safe behavior (uncomfortable, time-consuming, makes work difficult), management and peer failure to consequate the at-risk behavior in the past, and so on. Very often, when all these factors are considered, the behavior no longer appears "blatant and willful" but rather looks like a person trying to do the best job he/she can given the circumstances.

Those who draw the line need to be able to analyze the interaction between antecedents, consequences, and behavior (called "contingencies"). When such contingencies (established and maintained by management) are heavy contributors to an incident, it is inaccurate to call the performer's behavior "willful and blatant."

BLAME THE SYSTEM INSTEAD OF THE PEOPLE

It should be clear by now that we are recommending a philosophical shift, not just a tactical one. The philosophical shift is to view accidents, incidents, near misses, and at-risk behavior as *failures of risk management* rather than a function of bad individual choices. While this idea is not new (the nuclear industry talks about accidents being the result of *latent organizational weaknesses*), few organizations operate in ways consistent with this philosophy. An underlying belief remains that human error at the point of the accident is the cause of most accidents. This leads to a focus on fixing the behavior of the humans who make the errors. Dekker, James Reason, and others who write about managing risk argue that human error is a symptom, not a cause. As Reason says, "Errors... are shaped and provoked by upstream workplace and organizational factors."[5] Thus, it is not the human that needs fixing, but the system within which the human works. To be sure, the systems are created and managed by human behavior. This makes behavior analysis the ideal tool for improvement. But the target of the behavior change effort is very different. The target is to change the behavior of those who manage the systems so they change the systems to help prevent accidents.

Union criticism of behavior-based safety is often founded on the belief that BBS "blames the worker." While this may be true of some BBS processes, for those processes that have a solid foundation in behavior analysis, nothing could be further from the truth. A truly behavioral approach looks at the environmental contingencies (interactions of antecedents, behaviors, and consequences) within which behavior occurs and attempts to change behavior by changing those contingencies. To view an incident as a failure of management is to say that the organizational systems and contingencies have not been set up and/or managed well enough to prevent the incident.

Rather than blaming the worker, this approach seeks to identify conditions and events in the environment that encourage at-risk behavior. When possible, the systemic contingencies that encourage at-risk behavior are changed. This is the role of leadership in ADI's BBS process.

However, changing those contingencies is not always easy. Sometimes the change takes a long time, sometimes it is very expensive, and sometimes it is impossible. Changing workplace safety contingencies may involve purchasing new equipment, modifying existing equipment, changing the material flow or process, and/or changing organizational policies and individual managers. In those cases the focus should be on modifying the behavior of those in the line of fire to keep them as safe as possible given the circumstances. In such cases the contingencies are modified largely by the addition of peer feedback and reinforcement. This increase in consequences for the safe behavior offsets the systems consequences that inadvertently encourage at-risk behavior. For example, it is well known that one of the primary systemic root causes for people working without safety goggles is that the safety goggles fog up. The ideal systems solution is to supply goggles that do not fog up. Unfortunately, that proves to be difficult if not impossible. So what is the solution? Punishing people who don't wear safety goggles doesn't work. The best solution is to add feedback and positive reinforcement to make it more likely people will wear their goggles until it becomes habitual. That is what BBS usually does, but such incremental approaches should not be done without first looking for systemic solutions to safety problems.

PERFORM BEHAVIORAL INVESTIGATIONS

A typical incident investigation asks questions about training, process, procedures, equipment, and communication around hazards, but often fails to identify the more subtle organizational contingencies that contributed to an incident. Many possible contributors will surface if the right questions are asked. Some examples include overly complex processes, production pressures, peer pressure, level of fatigue, and lack of reinforcement for critical safe behaviors. The key is to understand the power of immediate and certain consequences. We believe that many of the most influential root causes are ignored or quickly dismissed because they seem small and unimportant. If they involve immediate and certain consequences, they are anything but unimportant.

Here are some questions that begin to get at the more subtle contingencies that often impact safe and at-risk behavior:

- How frequently does the at-risk behavior occur?
- Is the person at the center of the investigation the only person who engages in the behavior?
- What have supervisors and managers done in the past when the at-risk behavior has been observed?
- Equally importantly, what have they done when the desired safe behavior has been observed?

- What are the natural consequences for the safe and at-risk behaviors? Is the safe behavior time-consuming, uncomfortable, or does it make doing the work more difficult? Has anything been done to modify these consequences to make the safe behavior easier? If not, has positive reinforcement been increased to offset the negative consequences?

- Was skills training done to the point of fluency, or just mastery or awareness?

- What is the workload and what pressure is placed on workers either by management, peers, or self-imposition that might encourage risk taking? Will engaging in safe behavior be seen as interfering with meeting production or quality goals or expectations?

- What measures and consequences are in place for production, quality, or other business drivers? Might any of these inadvertently encourage risk taking?

- How do other workers (peers) contribute to the contingencies? Are they poorly trained? Are they also engaging in at-risk behavior? Do they give feedback when at-risk behavior is observed? Do they offer to help? Do they tease people who comply with safety rules? Do they reinforce safe behavior?

- Does the performer work alone? What contingencies are missing or added because he/she works alone?

- Are there incentives that might impact safety (for example, production incentives, quality incentives, safety incentives)?

- How do work procedures interact with safe behavior? Would following the procedure completely be overly difficult or time-consuming? If so, has positive reinforcement been increased to offset the negative consequences?

- Did the incident involve new or modified equipment? If so, were those who operate and maintain the equipment included in the design/decision-making process?

- Did the incident involve a new process? If so, were those who are to follow the process included in the development of the process?

- How strong is the general "sense of urgency" in the workplace? Are people positively reinforced for getting things done as soon as possible?

Careful analysis of all sources of contingencies and their contribution to at-risk behavior would add significantly to incident investigations. These contingencies should be the target of blame and the focus of accountability in moving forward.

USE THE PIC/NIC ANALYSIS® DURING INVESTIGATIONS

The PIC/NIC Analysis is a helpful tool for analyzing the contingencies that encourage or discourage targeted behaviors. When done with the performer in a context of trust and respect, it helps uncover the more subtle, but often most powerful of contingencies that drive at-risk behavior. Investigations almost always require several PIC/NIC Analyses to fully understand what happened.

The analysis usually starts with frontline employees who were involved in the incident. That first analysis usually points to the behavior of others as contributing to the behavior of the frontline performer. These analyses often cascade up and out through the organizational ranks helping shed light on how the behavior of others became part of the antecedent and consequence mix that resulted in the accident. (See Chapter 11 "Using Science to Understand At-Risk Behavior: PIC/NIC Analysis®" for an example of a PIC/NIC Analysis during investigations.)

PRACTICE FORWARD-LOOKING ACCOUNTABILITY

After an investigation has resulted in a good understanding of all root causes, the most critical task is to establish an accountability system for remediation of all root causes.

If equipment or conditions were found to be a contributor, then a plan should be developed for how those will be fixed.

If management systems such as incentive systems are found to be contributors, then a plan to change those systems should be created.

If engineering designs are found to be a root cause, then a plan should be developed to change the process for designing equipment.

If supervisory behavior or peer behavior was a contributor, then a plan should be developed to change those behaviors.

If the individual performer's behavior was a contributor, then a plan should be made to help him or her develop safer habits.

Keep in mind that behavior is at the heart of all of these root causes. Behavior change is required for equipment repair, for changing engineering procedures to include frontline input, for modifying incentive systems, and for improving how supervisors and peers interact with each other. Remember that plans change little unless consequences for the planned behaviors have been developed and delivered. Also remember that most of these behavior changes will take time; therefore, long-term follow-up is required. This is where effective accountability is required. Forward-looking accountability is a system to ensure these changes occur over the long term.

CREATE A CULTURE OF TRUST

As noted earlier, the use of punishment destroys the trust required to truly understand and remedy all the potential causes of accidents, but avoiding the use of punishment is not the only step required to build a culture of trust. Reporting at-risk behavior, near misses, or incidents should be actively and positively reinforced. Because reporting a mistake is naturally punishing (most of us do not like to admit when we did something wrong), it needs to be encouraged. Even reporting unsafe conditions, which has no significant natural punisher, should be positively reinforced. Any organization that has struggled with an ineffective near-miss reporting process has learned this.

When such a program is new, people will report near misses. Then they will watch to see what happens. In many cases what happens is . . . nothing. They don't get any feedback on their report and nothing changes because of the report. After a few such experiences, most people stop reporting. In other words, reporting undergoes extinction. Exemplary safety organizations get excited about reporting of near misses, unsafe conditions, and at-risk behavior because they know it presents them with the opportunity to better manage risk.

Of course, there is more to trust than just minimizing the use of punishment and positively reinforcing reporting of near misses. Trust is ultimately about relationships and is built over time as people interact and deliver consequences to each other. There are opportunities every day to build trust (or erode it) around safety. Trust is built when mistakes are treated as opportunities to learn. Trust is built when we follow through on what we learned in order to make sure it doesn't happen again. Trust is built when we stop blaming people who are trying to do a good job. Trust is built when management honestly assesses their role in at-risk behavior. Trust is built when all aspects of safe performance are positively reinforced.

Summary

To summarize, the danger of using punishment is that you create a culture of fear and mistrust and that results in people hiding their mistakes and not reporting them so that they and others can learn from them. The tremendous loss of valuable information that could be used to create a safer workforce is a tragedy. The irony is that punishment rarely has the desired impact anyway.

Yet, incidents must be dealt with. We have attempted to show that it is possible to respond to incidents in a behaviorally sound way that will result in lower probability of recurrence. While every incident is different and complex, consider this general approach, which summarizes some of our earlier recommendations:

- Approach the incident by assuming it is a failure of management, not the failure of an individual.

- Conduct an accident investigation and identify all root causes, including the behavioral root causes.

- Conduct a PIC/NIC Analysis of all identified at-risk behaviors (at all levels).

- Create a plan to remediate all possible root causes as soon as possible, including changes to management practices, organizational systems, et cetera. that may have contributed to at-risk behavior.

- Make the action plan public and highly visible; post it on a bulletin board, for example.

- Create a frequent and public accountability system to complete action items. Assign items to individuals and have those individuals report on progress as a part of regular meetings. Involve senior management in the accountability. Celebrate when action items are complete.

- When action items are complete, update the plan so all can see what has been accomplished and what is yet to be accomplished.

- As action items are completed, management should check back with the performers involved in the incident and/or others performing the same tasks to assess the probability of the same mistake occurring. In other words, have the action items truly changed the behavior?

- Build an ongoing process for checking that the root causes remain addressed.

- Remember, if managers are to change their behavior, positive reinforcement is required to sustain and maintain those changes.

---

[1]Dekker, S. (2007). Just culture: balancing safety and accountability. Ashgate Publishing Limited. Burlington, VT. P. 96.

[2]Dekker, S. (2007). Just culture: balancing safety and accountability. Ashgate Publishing Limited. Burlington, VT. P. 23.

[3]Sharpe, V. A. (2003). Promoting patient safety: An ethical basis for policy deliberation. Hastings Center Report Special Supplement, 33(5), S1 – S20.

[4]Dekker, S. (2007). Just culture: balancing safety and accountability. Ashgate Publishing Limited. Burlington, VT. P. 78.

[5]Reason, J.T. (1997). Managing the risks of organizational accidents. Ashgate Publishing Limited, Burlington VT. P. 126.

Seven Safety Practices that Waste Time and Money

Practice #6: Misunderstanding Near Misses

Learn from the mistakes of others—
you can't live long enough to make them all yourself.
– Martin Vanbee

In the last chapter we talked about punishing people who make mistakes, including near misses. In this chapter we are going to look at what a near miss is from a behavioral perspective.

We think the best definition of near miss is any deviation from a prescribed safety process. Whether in an office or a factory, there should be a prescribed way to produce a product, do office work, or deliver a service to a customer in a safe, efficient manner. Any deviation from that should be classified as a near miss. In the context of this definition, most near misses are not as dramatic as finding yourself swerving your vehicle to miss oncoming traffic at the last moment, and as such are usually what most people would consider "normal variance." Our position is that *there is no normal variance in safety.* Every deviation should be reported, observed, analyzed, and corrective action taken. We must sensitize employees to observe deviations in their own behavior and that of other employees. If you are a perfect employee and you notice that you tend to skip a step in a procedure, you would be quick to notify someone to observe your behavior, analyze the problem, and take action to help you eliminate that tendency.

In the punishment chapter we talked about why employees don't report their errors or those of peers. In a nutshell, no one minds someone looking over their shoulder to admire their work, but everyone minds someone looking over their shoulder to criticize their work.

When doing a good job is "what is expected" or "what you are paid to do" and the only time you hear from management is when you have made a mistake or done something wrong, it is highly unlikely that you will say, "Look at me! I did something wrong and I almost had an accident."

We have never worked in an organization that did not show an increase in the number of near misses after introducing a positive accountability safety process. Why do the near misses increase? They don't. Only the reporting of them increased. This tells you that previously reporting a near miss produced negative consequences for the employee. While managers in these organizations often deny that any negative consequences are associated with the reporting, the behavior indicates otherwise. Failure to report near misses is a characteristic of a negatively reinforcing and punishing culture.

In many airline crashes, the co-pilot knows the captain is making an error, but because of a culture of never correcting a more experienced person or questioning the authority of the captain, they fly to their deaths. A similar situation occurs in hospitals. Nurses often will not speak up when they see a physician making an error. The nurse will question his/her own knowledge before confronting the physician in the operating room or in front of the patient.

While such failures are often reported as pilot error or failure of the medical team members to exercise their responsibility, it is really a leadership failure. Leaders are responsible for creating the conditions under which such behavior occurs.

We are constantly amazed at how unsophisticated some senior managers are with respect to behavior. The management of one of our energy company clients always preached "individual responsibility." Senior managers claimed that not knowing about the occurrence of an error made it much worse. In written and oral communications managers frequently asserted that if employees made a mistake it was their individual responsibility to admit it. They related that such information could prevent an accident and allow management to take corrective action before the near miss turned into an expensive hit.

Unbelievably, when people admitted errors, they were fired or punished. The senior managers justified their actions because they said the near misses were serious and could potentially result in deaths and large economic losses. The net result of their actions was that few near misses were reported and, when errors were discovered, employees tended to blame others rather than accept "individual responsibility."

The Value Of Near Misses

Near misses provide valuable information about training, supervision, and teamwork. If a new employee has a near miss, it is probably an error of training. The employee may have been put on the job before reaching fluency, for example. If a seasoned employee has a near miss, it could be that supervision overuses negative reinforcement or team members tend to overlook, or cover, for the one who had the near miss. However, failure to *report* near misses is usually the result of poor management and supervisory practices.

Even though near misses provide a lot of diagnostic information about the workplace culture, you don't want a lot of them. As mentioned above, when instituting a high-performance safety culture, we routinely experience an increase in the number of near misses reported. Over time, the number will go down but will not go to zero. When the number goes to zero, negative consequences are usually present simply because whenever a large number of employees are engaging in tens of thousands of behaviors every day, there are almost always occurrences that almost caused an accident. However, when the reporting of a near miss is always the occasion to learn how to create a safer place to work, reporting such actions and events will almost always happen.

The Integrity Of Behavioral Chains

So what does a near miss tell us? From a behavioral perspective, it tells us that, for an experienced performer, the behavioral chain or habit is weak. The chain lacks integrity. The integrity of a behavior chain refers not only to an undivided or unbroken completeness of a performance but also to the tempo or rhythm of the performance as well. A herky-jerky performance with all steps completed correctly but where hesitations are common still lacks integrity.

A Northwest Airlines Flight 255 bound for John Wayne Airport in Santa Ana, California, originating in Detroit crashed on takeoff on August 16, 1987, killing 154 of 155 passengers and crew.

The NTSB probable-cause statement said, "The National Transportation Safety Board determines that the probable cause of the accident was the flight crew's failure to use the taxi checklist to ensure the flaps and slats were extended for takeoff. Contributing to the accident was the absence of electrical power to the airplane takeoff warning system which thus did not warn the flight crew that the airplane was not configured properly for takeoff. The reason for the absence of electrical power could not be determined."

While it was not a near miss (unfortunately it was a devastating hit), we suggest that there were many near misses in the history of these very experienced pilots. While we cannot be certain, the nature of the accident suggests a history of deviations from the checklist routine.

Northwest Airline investigators claimed they heard on the voice recorder the command, "Flaps" and the response, "Flaps extended," but the NTSB report stated that there was no mention of flaps or slats. If the pilot and co-pilot did indeed announce the flaps and slats command and response, it is obvious that they did not check. One of the reasons they didn't check may lie in the fact that they were assigned to the wrong runway and in the confusion failed to check. After all, they had probably never taken off with the flaps and slats in the wrong position in almost 50 years of cumulative experience. It is possible that on numerous occasions in all those years they could have failed to check and still had perfect take-offs. This history would have weakened the integrity of the checklist procedure making them distracted by the confusion in runway assignment—a serious error on the part of the air controller. In addition, the fact that they were distracted by the error, as serious as it was, also suggests that their routine was not fluent.

How could it be that experienced pilots were not fluent in something as mundane as a checklist? In any behavior chain, if an item is missed or performed incorrectly, the chain is modified ever so slightly.

Each time that the chain is repeated successfully without incident even though the behaviors were not as prescribed, a new chain or habit is being formed and strengthened. Each time a step is skipped, executed poorly, or not in a timely way, the performers are inadvertently getting positive reinforcement (PICs) for deviations from the prescribed procedure. Casual observation of employee performance will often not pick up the small deviations in behavior that indicate that the desired behavior chain is not receiving sufficient reinforcement.

The problem is that even with the modified (deviated) chain, the results are often still achieved. Everyone as a child has had the experience of having your shoelaces come untied. The last step you were taught, if you had a good teacher, was to pull the laces tight. Unfortunately for you, even though you failed to pull them tight, they remained tied for a period of time. Later, you might have tripped on them, had your shoe come off, or had to stop in the middle of a game (to the dismay of your teammates) to re-tie your laces. Everybody can remember some children who usually had shoelaces flopping as they walked or ran. The parents of such children were probably frustrated by their children's continual inability to tie their shoes correctly and may have eventually resorted to something we have all heard, "I guess you will just have to learn the hard way."

As adults, we learn the same way. It is easy to drift into ways of responding that are variations of what we were taught. This drift creates vulnerability to distraction, fatigue, and unusual circumstances. Airline pilots have to go on check rides to inspect the integrity of the behavioral chains in the cockpit. Industrial safety, with few exceptions, does little to ensure the integrity of behavior chains.

Ensuring The Integrity Of A Behavior Chain

As work becomes more complex, behavior chains are often longer. Long behavior chains increase the probability that a step can be left out, or modified, without affecting the outcome. Every time this happens, the behavior chain is corrupted. Unfortunately, the changes are so subtle that they go unnoticed the vast majority of the time until a failure of some kind occurs.

In the book, *The Checklist Manifesto*,[1] Dr. Atul Gawande describes many such failures in hospital operating rooms. The number of steps in even simple operations has increased through technology and new techniques. Checklists are used to make sure that all of the steps are completed in the right sequence. As we will discuss in the next chapter, the checklist is a valuable job aid, but its value is determined by how well each of the steps are executed and that is determined by how well feedback and reinforcement are provided for the behaviors in the chains.

Small Deviations Matter

Announcers of the televised Olympics point out deviations from prescribed routines, such as those in figure skating, that are so small the average viewer cannot see them, even in slow motion. Highly successful athletic coaches are usually very skilled at behavioral observation. Because they have seen many thousands of performances, they can detect small changes that when shown to the athlete result in quick change in the behavior and subsequent results. In golf, the position of the hands on the club, or even of one finger, can make a huge difference in ball flight. In figure skating, a coach is able to see that the landing from a triple axel is off by only a very few degrees. In baseball, a coach may change the position of the batter's feet for a better chance at hitting the ball. In all of these cases the coaches, at least the good ones, cause the athlete to practice a correct behavior chain multiple times until he/she develops muscle memory for the new chain. This may require 300 or more repetitions.[2]

Management Response To A Near Miss

Despite acknowledging the value of near-miss information, management's faulty response to near-miss reporting often results in a decrease in reporting, not an increase. The earlier example of the energy company that wanted better "individual accountability," highlights at least two problems with management actions. The first is they punished the behavior they wanted. It didn't take long for other employees to see that it was not to their benefit to report near misses. The second is that punishing a behavior doesn't correct it. As noted in previous chapters, punishment only stops behavior. In addition, since the punishment is usually delayed, often occurring days after the incident, it is unclear which behaviors were actually impacted. Often the behaviors changed are those that help the performer avoid getting caught in an unsafe act. (Learn more at www.aubreydanielsblog.com, "Safety Leadership: Who's Accountable?")

What to Do Instead

 As noted previously, a near miss should be viewed as a failure of management, not a bad individual choice.

Careful analysis of the conditions under which the near miss occurred will identify changes that should be made in order to reduce future near misses and accidents. Attend to the following:

Sweat the small stuff. There is no such thing as normal variance in safety. Flag even the smallest variance in procedures and processes. Variance in the time it takes to complete steps or tasks is an early warning sign of later errors and near misses.

Develop systems to evaluate near misses. Catalog near misses to identify any patterns in location, time of day, time of month, product, supervisors, and individuals. Each of these will help identify causes and remedies that are not proximate to the near miss. Keep in mind this system can be and should be a joint effort with the hourly population. Over time each organization can develop an efficient process that gleans valuable information relatively quickly.

Correcting the Near Miss. Musicians know more about how to correct a mistake than do most managers. When a mistake is made while practicing a musical composition, the best conductors spend a minute examining the error to make sure that the musician(s) can execute the correct note(s), and then he/she "takes it from the top." Since the conductor is interested in a flawless performance he/she is interested in having chains performed correctly many times. Practicing only the behavior that is a problem often teaches the musician to make the error and then to correct it.

Correcting a near miss requires several steps:

1. Look before you listen. Go to observe the behavior in question. Ask the performer to show you what happened and look both at the behavior and the context within which it occurred.

2. Do a dry-run. Have the performer execute the behavior correctly to ensure it is not a training problem.

3. Start at the top. Ask the performer to demonstrate the

entire sequence of behaviors from start to finish, if possible. See where problems exist—hesitations, barriers, systemic flaws.

4. Correct any problems of training, fix any physical barriers, and adjust any management-controlled systems that contributed to the problem.

5. Start at the top again. Repeat the entire chain multiple times until the correct behavior is executed correctly and fluently several times in a row.

6. If possible, repeat the chain under a variety of conditions that represent the variability likely to be experienced while doing the work. Simulate high-pressure circumstances, situations where something doesn't work correctly, or where others don't do what they are supposed to do. These "variable fire drills" are unlikely to cover all possible upset conditions, but the practice will increase the probability the correct chain will persist under a variety of circumstances.

By working on a no-fault, no-blame process for dealing with near misses, valuable lessons can be learned and the focus can be on helping employees develop the habits that preclude errors.

---

[1]Gawande, A., (2009) *The Checklist Manifesto: How to Get Things Right.* New York: Metropolitan Books.

[2]Schmidt, R. A., (1991) *Motor Learning & Performance: From Principles to Practice.* Human Kinetics Books. Champaign, Ill. p.215.

Seven Safety Practices that Waste Time and Money

Practice #7: Thinking That Checklists Change Behavior

A checklist is nothing more than a job aid.
– Aubrey Daniels

The use of checklists has a long history within organizations, and safety systems typically include many checklists: inspection checklists, procedure checklists, machine shut-down and start-up checklists. Checklists are an integral part of behavior-based safety processes. Lists of safe behaviors are used to measure the degree to which groups or individuals are working safely.

More recently, there has been a rebirth of interest in the use of checklists, particularly in the medical field, to reduce medical errors. With the publication of *The Checklist Manifesto,*[1] Dr. Atul Gawande and Dr. Peter Pronovost have appeared frequently on TV and other media outlets discussing their use of checklists to prevent medical errors. Dr. Gawande's book recounts his interesting quest to make surgery safer. He finds the checklist to be a common tool used by a variety of industries, and his research on the impact of using the checklist during surgery is impressive. Using checklists, the rate of major complications for surgical patients in eight hospitals dropped by 36 percent and deaths fell by 47 percent. Infections fell by almost half. Michigan hospitals began implementing Dr. Provonost's checklists in Intensive Care Units (ICUs) in 2003. Within three months, hospital-acquired infections in a typical ICU in the state fell from 2.7 per 1000 patients to zero and sustained this spectacular result for years. The credit is given to checklists.

We have recommended the use of checklists to clients for the 30-plus years we have been in business. Checklists are great tools that assist with changing and managing behavior. A carefully composed checklist can serve as (1) an antecedent (providing instructions on what to do and the correct order to do it), (2) a measurement tool (for tracking that tasks were done), and (3) a feedback mechanism (the performer can see progress by checking items off the list). However, checklists do not provide consequences. In order to produce lasting change, they must be used in conjunction with consequences.

As noted in the inset ("A Checklist Never Saved a Life," page 98), failure to acknowledge that the consequences surrounding the checklist bring about behavior change leads to misuse of the checklist. Checklists alone do not improve safety. There are countless examples of safety checklists of various sorts being pencil-whipped. The only behavior that changes when a checklist isn't linked to meaningful consequences is the behavior of filling out the checklist.

The assumption that the checklist is the cause of behavior change is also seen in BBS. Because behavior observation checklists are an integral part of BBS, organizations sometimes focus on the checklists to the exclusion of consequences. Feedback and reinforcement change behavior, not observation checklists.

Why Do Companies Do It?

Checklists are recommended tools in a variety of settings. Some checklists are required by OSHA and other regulating bodies.

Sometimes performers themselves create checklists to help them do their jobs. Often the reinforcement associated with the checklist is not obvious so it is easy to assume the checklist is changing behavior rather than consequences. For example, "to do" lists that many of us create in our work and home life appear not to have reinforcement associated with them. There is no one saying "good job" when we check an item off of our list. However, for those that use these checklists regularly, reinforcement takes place. The fact that another task is completed is usually the primary reinforcer. Seeing more and more items get checked off is cumulatively reinforcing. It is not the checklist that is driving the behavior; it is the evidence (consequence) of making progress. But these reinforcers don't exist for all checklists or for all people using the same checklist.

Sometimes people assume a checklist is all that is required to change behavior because a newly instituted checklist will often result in temporary behavior change. This is a predictable effect of an antecedent—short-term behavior change. That short-term change is reinforcing to the user and can lead to the assumption that the checklist solved the problem.

What To Do Instead

Again, we want to be clear. Checklists are a great tool. Clear antecedents are incredibly important in supporting safe behavior. The point is that checklists are most effective when combined with consequences.

Do the following to make checklists maximally effective:

- Make sure that a check on the checklist represents behavior or accomplishments that have been completed in the proper and safe way.

- Make sure that items on the list are observed apart from the checklist to ensure quality and safety of the performance.

- Finally, ensure that the use of checklists is paired with positive reinforcement. When following a checklist leads to reinforcement, the checklist becomes an important tool for developing sound behavior chains and producing long-lasting change.

> ### A Checklist Never Saved a Life from
> ### www.aubreydanielsblog.com

When I read *The Checklist Manifesto: How to Get Things Right* by Atul Gawande, three thoughts occurred to me. The first was a joke someone sent me in an e-mail last week: A man is recovering from surgery when the surgical nurse appears and asks him how he is feeling. "I'm O. K. but I didn't like the four letter-word the doctor used in surgery," he answered. "What did he say?" asked the nurse. "He said, 'Oops!'" the patient replied.

The second thought was a quote by Norman Cousins, long-time editor of the *Saturday Review.* After being hospitalized with a rare disease, he said, "I soon realized a hospital was no place for a person with a serious illness." I would certainly not recommend reading *The Checklist Manifesto* to anyone scheduled for surgery. Even with the impressive gains that Gawande reports for hospitals using the checklist methodology, a lot of dangerous things still happen in the surgery suite.

My third thought was that team leadership is not a highly developed skill of most surgeons. Dr. Gawande defines a checklist as a way of organizing that empowers people at all levels to put their best knowledge to use, communicate at crucial points, and get things done.

I can tell you with certainty that a checklist does not empower anyone to take critical actions required to avert disaster or even to question the actions of other team members. The history of aviation is filled with examples of co-pilots who will not question the captain of an airliner even when he observes the captain making a fatal error. Dr. Gawande gives an example of one such instance in his book. The Tenerife airport disaster in 1977 occurred when a KLM plane crashed into a Pan AM flight that was still on the runway. Twice, when questioned (once by the co-pilot and once by the engineer) whether clearance to takeoff had been given, the captain ignored them and continued the takeoff resulting in the deadliest airline disaster in history.

Is it possible that even with checklists, a nurse or resident would fail to correct an action of the surgeon? Dr. Gawande says that Brian Sexton, a Johns Hopkins psychologist, found that 25 percent of surgeons believe that junior team members should not question the decisions of a senior practitioner.

Dr. Pronovost, who started the checklist idea at Johns Hopkins, said in an interview in *The New York Times*, "When I began working on this, I looked at the liability claims of events that could have killed a patient or that did, at several hospitals—including Johns Hopkins. I asked, 'In how many of these sentinel events did someone know something was wrong and didn't speak up, or spoke up and wasn't heard?'"

He went on to say, "Even I, a doctor, have experienced this. Once, during a surgery, I was administering anesthesia and I could see the patient was developing the classic signs of a life-threatening allergic reaction. I said to the surgeon, 'I think this is a latex allergy, please go change your gloves.' 'It's not!' the surgeon insisted, refusing to change his gloves. So I said, 'Help me understand how you're seeing this. If I'm wrong, all I am is wrong. But if you're wrong, you'll kill the patient.' All communication broke down. I couldn't let the patient die because the surgeon and I weren't connecting. So I asked the scrub nurse to phone the dean of the medical school, who I knew would back me up. As she was about to call, the surgeon cursed me and finally pulled off the latex gloves."

I ask you, what nurse, resident, or anesthesiologist would have done what Dr. Pronovost did? Consider the power of position, of rules of conduct, and of histories of reinforcement for "showing respect" in the medical field.

A checklist is nothing more than a job aid. We have helped people in all kinds of organizations develop and use them for almost 40 years, so I don't dispute their value. While a checklist adds value to almost any process, the real value is determined by what happens to the behaviors surrounding actions required by the checklist. Dr. Gawande fails to realize that he is actually introducing a new process into the surgery room. The checklist only gives him access to opportunities for implementing his process. It does not empower anyone: the checklist is just another of the tools that assist medical personnel in completing the job in the best possible manner. If the checklist did what he purports, "empowers people at all levels to put their best knowledge to use, communicate at crucial points, and get things done," surgery omissions and other errors could be immediately addressed by giving all who enter the operating room a checklist to follow.

In all the successful cases reported in the book, the *behavioral consequences* were changed for the surgery team. The task of constructing a checklist forces the surgery team to pinpoint tasks, roles, and responsibilities more specifically than before. In addition, spending time together deciding on content and conduct of the checklist changes the working relationships.

Most of the places where good results were obtained were places where the surgeon personally introduced the checklist. This generally produces a different reception of a new process than if someone from Training or Infection Control had done so. By having the surgeon introduce the checklist, at a minimum it implies, even if it is not said, that "I want us to follow this procedure during surgery." This immediately changes the consequences of speaking up or calling attention to items on the checklist. It actually increases the probability that the behavior of stopping the process or calling attention to a problem will be positively reinforced rather than punished. Simply taking time to introduce the members of the team to one another before starting the operation, which is not always done, produces expectations in many people that this is a different kind of operating room than they have encountered in the past. Such a seemingly inconsequential step can change the consequences for team-member behavior. *Consequences* change behavior; checklists don't.

In spite of everything I have written, I would not discourage any hospital from using anything that produces a safer experience for the patient. It is just that when surgeons, and hospital personnel in general, understand the behavior-change process, results will be even better and sustained longer than when they think that major improvements can be made and sustained just by creating a checklist. People doing the right things, at the right time, and in the right way are the true life-saving factors.

In my opinion, hospitals, and medical schools need to focus on recruiting physicians who are team players and on training the existing ones in the science of behavior. For a good process to work anywhere, every team member must feel comfortable speaking up when things don't seem right. Team members should be positively reinforced for questioning the actions of any team member, not punished or ignored. While Dr. Gawande might have thought he was saying these things, the focus of his book is on the checklist not the team-member interactions—interactions that are the *real* key to creating and sustaining change.

Some of the key actions that must be reinforced to produce the best outcome in the operating room are as follows:

1. Introducing the team

2. Reviewing the surgery and all relevant facts about the patient

3. Verbally thanking anyone who questions a procedure or calls attention to a step on the checklist

4. Conducting a post-surgery review of what went right and what was problematic

5. Calling attention to good performance of individual team members during surgery

6. Modifying the checklist as necessary so that the next team will be able to perform even better than before

In the final analysis, following a checklist requires behavior. Whether it will be done well, poorly, or at all depends on the consequences surrounding the use of checklists. Without a good understanding of the behavioral process, checklists have an uncertain future in the operating room or any other part of medical practice. With an understanding and correct application of the behavioral process when using checklists, more lives could be saved.

---

[1]Gawande, A. *The Checklist Manifesto: How to get things right.* Metropolitan Books: New York, P. 154.

PART THREE

Effective Safety Leadership

Up to this point we have shared some safety practices that don't work and offered suggestions on what to do instead. Now we will make some broader recommendations around the steps that leaders can take to create a high-performance safety culture.

We have chosen to focus on the following four leader activities because in our experience these represent the biggest opportunities for improvement. We call these the four pillars of safety leadership:

1. Relationship development

2. Using science to understand at-risk behavior

3. Maintaining a safe physical environment

4. Creating systems that encourage safe behavior

We will provide details of each in the following chapters.

The leader's role is to establish the conditions under which all performers will choose to execute the mission, vision, and values of the organization.
– Aubrey Daniels & James Daniels, *Measure of a Leader*

CHAPTER 10

Effective Safety Leadership

Relationship Development

The day soldiers stop bringing you their problems
is the day you have stopped leading them.
They have either lost confidence that you can help them
or concluded that you do not care.
Either case is a failure of leadership.
– Colin Powell

As we have noted several times, a high-performing safety culture requires that all employees be engaged and willing to discuss, address, and change organizational weaknesses that might lead to accidents. In a previous chapter, we discussed the role of punishment in preventing openness around safety. Fear of punishment leads people to avoid speaking openly and reduces engagement. But there is more to developing good relationships than simply avoiding the use of punishment. Excellence in safety requires discretionary effort—effort that goes above and beyond that which is required and beyond that generated by lack of fear.

A central tenet of all the work we do at ADI is that effective positive reinforcement leads to discretionary effort. This is as true in safety as in other parts of business. But the simplicity of the statement conceals the complexity of making it work. Discretionary effort is not generated by making a few positive comments to direct reports now and then. Positive reinforcement can only be maximally effective in the context of a good relationship. Therefore, discretionary effort will increase as personal relationships improve.

What's Love Got To Do With It?

The following behavior analysis of relationships at work helps us understand why this is so. Relationships are important because they alter reinforcer effectiveness. In other words, relationships set the context for how effective reinforcement will be.

By now readers know that the impact of a consequence is modified by its immediacy and certainty. The other way the impact of a consequence can be modified is through a Motivating Operation (MO). As noted earlier, an MO is an event that alters the effectiveness of a reinforcer or punisher. For example, food is more reinforcing the hungrier we are. Hunger (or food deprivation) is an MO. Hunger establishes food as a reinforcer. Relationships are also MOs. A positive statement such as "I appreciate your input on this" or "I was impressed by how you handled that meeting" will have greater or lesser reinforcing value depending on the relationship we have with the deliverer. From a boss that we like, trust, and respect, such comments will likely be powerful reinforcers; from a boss that we do not like, trust, or respect, the comments will likely have little positive impact and may likely have a negative impact. Thus, your ability to accelerate business performance through the use of positive reinforcement is to a large extent a function of your relationships with those whom you attempt to reinforce.

Telling a person that she is doing a good job assumes that she cares about what you think. If she doesn't like you, it's unlikely that any attempts at positive reinforcement, coming from you at least, will work.

The impact of relationships on safety is a phenomenon noted by many. For example, most managers have observed that supervisors who have good relationships with their crews tend to have better results in safety and everything else. The opposite is true as well. Bad relationships usually lead to weaker performance in safety and everything else. (Learn more at www.aubreydanielsblog.com, "Words, Just Words.")

Another piece of evidence is the fact that participation in safety programs is sometimes used as a bargaining chip during difficult union/management negotiations. In the context of poor relationships people are unlikely to participate in voluntary programs, get involved in improvement efforts, or become fully engaged in safety. In other words, they won't give discretionary effort.

It is our position that managers and supervisors are paid to be well-liked. While we have had many managers disagree and misunderstand that statement, when you fully understand the science of behavior, you understand this logic. An effective leader is one who has willing followers. When someone follows because he feels pressured, obligated, or threatened if he doesn't follow, it is certain that you will not get his best work. Therefore, good relationships are primary.

Many people have been promoted into management or supervisory positions because of their technical expertise or years of on-the-job experience. This is a huge mistake. The first screening that management candidates should have to pass is this: Do people like them? Do people want to be around them? Likeability is the first and most important criterion in selecting managers and supervisors. Expertise and experience take a subordinate role to being able to form positive relationships with others. It is not that expertise and experience are unimportant, just not the *most* important. As the economist, actor, and author, Ben Stein aptly said, "If people are failures at relationships, they are likely failures at everything."

The irony of this is that if you try to be liked, you probably won't be liked. We have all known people who "try too hard" to be liked. This behavior backfires most of the time. Managers and supervisors who are always positive, always say yes, and never set limits on behavior, are not only not liked, they are ineffective.

Being well-liked does not require being a finalist for a personality contest. We have known many shy, serious or reserved people who have been well-liked by their direct reports and thus have been successful.

People who demonstrate genuine interest and concern in the success of others are most often liked, respected, and appreciated. An effective leader is one whose reinforcement comes from helping followers become successful at whatever they are tasked to do. Making employees successful should occupy most of a safety leader's time, because making your employees successful is what leaders are paid to do.

Building Better Relationships

So what is a "good relationship"? The best definition is functional. A good relationship is one that works; one in which both parties work successfully together. Characteristically, in good manager/employee relationships, people volunteer to take on assignments, tasks, and responsibilities that are not formally a part of their job.

The structural characteristics of good relationships can vary greatly. Some good relationships are characterized by polite behavior and obvious common, positive regard. Other relationships are short on social niceties and long on mutual sarcasm, humor, and ribbing. The ultimate test of the relationship is in the answer to the question, "Does it generate discretionary effort?"

Despite the variety, we have found some consistencies in effective management/employee relationships in our work with clients over the last 30-plus years. We present these behaviors in the form of a checklist that the reader can use to improve relationships with direct reports.

BEST PRACTICES FOR BUILDING EFFECTIVE RELATIONSHIPS AROUND SAFETY

Set clear expectations.

Use pinpointing to ensure clarity of expectations; avoid assumptions, and ask recipient(s) to state an understanding of the expectations.

Listen.

Use active listening skills such as maintaining eye contact, using appropriate facial expressions, paraphrasing, and asking questions to demonstrate understanding. Avoid looking at or using computers and smart phones when others are talking to you.

Acknowledge good work, not just mistakes/problems.

Track the nature of your interactions. Good leaders maintain a higher ratio of positive to constructive comments/discussions.

Ask questions to understand problems/issues.

Avoid jumping to conclusions. There is always more to every story. Ask questions to uncover the details.

Ask for feedback about your own effectiveness and areas for improvement.

Seek detailed information about what you do well and what you need to do differently to be more effective.

Demonstrate that you are listening and working to improve your own actions.

Avoid blame.

Remember that people's behavior makes sense to them. Find out what the environmental contingencies were that lead to undesired behavior.

Respond fairly to incidents (safety and other types).

Better incident investigations will lead to fair responses. (see "Using Science to Understand At-Risk Behavior: PIC/NIC Analysis®," Chapter 11)

Admit when you make mistakes.

Acknowledging your own mistakes helps establish that mistakes are expected and that learning from them is critical.

Solicit input and opinions from direct reports.

Asking for input and advice will not only lead to better solutions, but in many cases, it also demonstrates respect.

Follow through on commitments.

Use whatever memory devices you need to be sure to do what you say you will do. Consistent follow-through is essential for building trust and respect.

Stand up for direct reports; "go to bat" for them.

Verbally promote direct reports and share their successes with others. In addition, acknowledge some responsibility when direct reports make mistakes.

Remove roadblocks in order to set direct reports up for success.

We have said before that the number-one job of management is to make direct reports successful. Analyze what gets in their way (the PIC/NIC Analysis is a great tool for this) and do what you can to remove obstacles.

Provide feedback that helps direct reports improve.

Pinpointed, timely feedback is most helpful. Don't save feedback for annual appraisals or even monthly one-on-one meetings; just-in-time feedback is the most effective.

Demonstrate that you trust direct reports.

Give employees appropriate responsibilities and avoid micromanaging. Tell them you trust them, when appropriate, and reinforce trustworthy behaviors.

Treat direct reports like people, not just employees.

Make a point to greet direct reports at the start of the shift (when possible); show an interest in their lives outside of work, and demonstrate concern and consideration.

A more in-depth discussion of the important topic of building management/employee relationships is beyond the scope of this book; indeed the topic is worthy of its own book. For practical ways to measure how well you are doing at developing relationships with others, see *Measure of a Leader*[1] by Daniels and Daniels. Our objective in this book is to emphasize the importance of building good relationships at work as a foundation for improving safety. Safety improves through the behavior of people. Employee behavior is driven by clarity of expectations but most importantly by consequences. Consequences are delivered more effectively in the context of good relationships. Thus, deliberate effort to build good relationships is worthy of time and attention.

---

[1] Daniels, A.C. & Daniels, J.E. (2005) *Measure of a Leader: An Actionable Formula for Legendary Leadership.* Atlanta: Performance Management Publications.

CHAPTER 11

Effective Safety Leadership

Using Science to Understand At-Risk Behavior: PIC/NIC Analysis®

*In all science, error precedes truth, and it is better
that it should go first than last.*
– Hugh Walpole

Our discussion of safety practices that waste time and money may have left the reader wondering why so many smart people use practices that only appear to improve safety. Why don't people know better?

In attempts to solve real problems, humans have a history of using and persisting with less-than-effective solutions until science has a chance to catch up and provide a more thorough understanding of the problem, which leads to more effective solutions. An old, but vivid example is the use of leeches in medicine. A better understanding of causes of disease showed the folly of this ineffective practice and paved the way for more effective ones.

Medical history is filled with examples where treatment was instituted with only partial success or tragic results until science demonstrated better knowledge of controlling variables. In the past, for example, British sailors were often referred to as *limeys*, a slang term often used as a pejorative. The term is believed to come from the Royal Navy and Merchant Navy practice of supplying lime juice to the sailors to prevent scurvy during long sea voyages when their diets were low in fresh foods. The benefits of citrus juice were well known at the time thanks to the experiments of surgeon James Lind who studied the effects of citrus on scurvy in 1747. While lemons were originally used, limes replaced lemons because they were readily available from the British Caribbean colonies.

Lemon juice was reintroduced after scurvy again became a problem due to the insufficient amount of vitamin C in limes. Even though the limes worked in some situations, later data indicated that lemons offered a better solution.

Interestingly, even though Dr. Lind demonstrated the link between citrus and scurvy in 1747 in what was considered the first clinical trial in medicine, his findings were not implemented until 1795, 48 years later. It is not unusual that scientific knowledge preceeds actual practice by many years.

Likewise, many safety practices have shown some effect in reducing accidents and injuries even though they have not been subjected to scientific scrutiny. We suggest that by replacing common sense with scientific fact, we will be able to point leaders to solutions that have a higher probability of sustained success.

In this chapter we will elaborate on the use of the PIC/NIC Analysis in incident investigations. In the following chapter we will analyze the rare error from a scientific perspective.

PIC/NIC Analysis

As we have stated several times, the PIC/NIC Analysis is an extremely useful tool to better understand at-risk behavior occurring at any level of the organization. The PIC/NIC Analysis can help us understand the systemic consequences which encourage bypassing a safety procedure, failing to wear PPE, or standing in the line of fire. It can help us understand why a supervisor focuses almost exclusively on correcting at-risk behavior and ignores safe behavior.

It can help us understand why an executive, who claims safety is a core value, does not behave in ways consistent with that value.

As noted earlier, blame and the negative consequences that usually follow have no part in a high-performance safety culture. Unfortunately, superficial analyses of incidents and at-risk behaviors usually lead to blame since the accident has already occurred and assigning blame seems like a natural thing to do. However, these analyses rarely get to the real causes of the behaviors since they don't include a systematic analysis of the antecedents and consequences associated with the behaviors. The PIC/NIC Analysis is a tool that provides such an analysis and thus provides a unique perspective on causes and accountability. By illuminating the multiple causes of at-risk behavior, blame is replaced with problem solving, because the real causes become more obvious. Inevitably, this process results in increased respect for all parties and a sense of fairness that feeds better relationships.

A Sample PIC/NIC Analysis

Following is a PIC/NIC Analysis of the at-risk behavior of a mechanic in a food manufacturing facility. The at-risk behavior is "skipping a step of the lock-out/tag-out (LOTO) procedure." In many companies, failure to fully lock-out and tag-out is cause for termination. Because it can lead to serious injury, destruction of property and/or death, it is not a behavior to take lightly. Since the mechanic working on the equipment that is not locked and tagged is most often the one who will get hurt or killed, it seems illogical for anyone to fail to complete this procedure. In reality this behavior (of failing to lock-out/tag-out equipment) is very common (although management usually is not aware of this fact).

In most cases, good training and clear Standard Operating Procedures (SOPs) are in place. Thus it is common to blame the mechanic for this behavior, because "He was trained and he knew better." This PIC/NIC Analysis (based on real cases) shows that there is often much more to the story.

At-Risk Behavior

Skipping steps of the LOTO procedure

Antecedents	Consequences	P/N	I/F	C/U
Broken equipment SOPs	Fix equipment faster	P	I	C
Training Sense of urgency, because if equipment isn't fixed the food product in process will be spoiled	Save product	P	I	C
	Avoid being teased by peers	P	I	C
Recent history of negative feedback to maintenance because of poor machine reliability	Avoid having to change shoes and wash hands	P	I	C
History of peers teasing mechanics who can't fix equipment fast enough to avoid product loss	Improve maintenance image	P	F	U
Engineering designs make LOTO difficult	Less hassle (engineering design makes doing LOTO a hassle)	P	I	C
Quality department instituted recent food-hygiene zones that require change of shoes, and hand washing prior to entry	Praise from supervisor for fixing it fast	P	I	U
	Can move on to another job more quickly	P	F	C
Short-staffed in maintenance department	Could get hurt	N	I	U
Did it before and nothing bad happened	Get disciplined/terminated	N	F	U

Review of the PIC/NIC Analysis should begin with the antecedents. In this case there are seven antecedents that prompt skipping steps in the process. The SOP and training are the only two that prompt the safe behavior and neither of these are likely to have happened just prior to the behavior. Research has shown that antecedents that occur just prior to the behavior have the biggest impact on behavior.

Looking at the consequences (and their classifications) shows the same bias toward the at-risk behavior. Given the number of possible positive, immediate, and certain consequences, it should be no surprise this behavior occurs. Many consequences encourage this at-risk behavior. Looking closely at the consequences listed shows that the mechanic was not just "being lazy" or "willfully violating rules," but rather was trying to get the job done quickly in order to avoid loss of food product and to move on to additional work. Furthermore, several obstacles to the safe behavior are within management control. This PIC/NIC Analysis, like most, leads to further questions and additional PIC/NICs on others' behavior. Some questions we would ask to further investigate this at-risk behavior include the following:

- Why was the maintenance department short-staffed and what was being done about it? In one such case, further inspection uncovered the staffing problem was partially due to a test created for screening applicants that was so difficult it was almost impossible to pass, thus making it extremely difficult to hire more mechanics (a management issue).

- Was it made clear to the mechanic that lost product was okay and he would not be blamed if he took the time to fully lock-out and tag-out (and all other safe behaviors) or was the communication focused on getting the equipment fixed as quickly as possible? In our experience, management believes they have communicated clearly that it is okay to shut down production, lose product, and so on, in the name of safety. To act as an effective antecedent, this is a message that should be repeated often, particularly during upsets and emergencies.

- What, if anything, was done to stop the teasing by peers? Such teasing may appear harmless, but it often contributes to at-risk behavior and should be systematically dealt with.

- Why did the engineering department not work with mechanics when designing the equipment to ensure it was not overly cumbersome to work safely on it? Why did the quality department not work with mechanics when designing the hygiene zones to ensure they were not making safe and timely repair more difficult?

- What had management done in the past when LOTO was not complete and when it was completed thoroughly?

Each of these questions may lead to a PIC/NIC Analysis of the behavior of the person(s) in question—management, the engineers, the quality staff—the ultimate goal being to make systemic and behavioral modifications whenever necessary to better support the safety behavior of the mechanics. In some cases, changes can be made to the antecedents to better prompt the desired behavior. In almost all cases, the solutions lie in changes to the consequences.

A second analysis, this time focused on the safe behavior, is often helpful. While analyzing the safe behavior can often produce a mirror-image analysis (see next page), the analysis often illuminates other actionable areas. In some cases, the missing elements are the most telling.

Note the absence of antecedents from the supervisor. Would a well-timed reminder to the mechanic that following all steps of LOTO is critical and that if it means losing product that is okay have helped to increase the likelihood of the safe behavior? Would offers to assist make a difference? How about checking in with mechanics in the middle of the process and reinforcing the following of procedure?

In our experience, in helping clients conduct PIC/NIC Analyses after an incident, we often find long-standing consequences that contribute to the at-risk behaviors. We also often find unique antecedents that clearly contribute to the behavior. As noted earlier in the book, accidents usually happen when multiple variables come together. Unusual circumstances (bad weather, breakdown of equipment, being temporarily short-staffed, the presence of contractors, et cetera) are all antecedents that can influence the decisions of performers in the moment. Recognizing the influence of such unusual antecedents and working to offset them (with antecedents prompting the safe behavior and beefed-up consequences for safe behavior) can help avoid incidents. In fact, one of the behaviors we encourage our clients to develop into a habit is doing a quick hazard analysis before beginning an infrequent job or a job that is unusual in any way. By stopping and reviewing the hazards and identifying behaviors to avoid the hazards, unusual circumstances don't turn into incidents.

Safe Behavior
Following all steps of the LOTO procedure

Antecedents	Consequences	P/N	I/F	C/U
Broken equipment	Avoid getting hurt	P	I	U
SOPs				
Training	Avoid discipline/ terminate for not following procedure	P	F	U
Sense of urgency, because if equipment isn't fixed the food product in process will be spoiled				
	Fix equipment more slowly	N	I	C
Recent history of negative feedback to maintenance because of poor machine reliability	Product may be lost	N	I	C
History of peers teasing mechanics who can't fix equipment fast enough to avoid product loss	Teased by peers	N	I	C
	Have to change shoes and wash hands	N	I	C
Engineering designs make LOTO difficult	Contribute to image of maintenance being slow	N	F	U
Quality department instituted recent food-hygiene zones that require change of shoes, and hand washing prior to entry	More hassle (engineering design makes doing LOTO a hassle)	N	I	C
Short-staffed in maintenance department	Pressure from supervisor for taking so long	N	I	U
Skipped it before and nothing bad happened	Can't move on to another job quickly	N	F	C

The PIC/NIC Analysis will not provide all the answers, but it will stimulate the right questions and facilitate discussions that lead to preventative action for the future.

CHAPTER 12

Effective Safety Leadership

Using Science to Understand At-Risk Behavior: The Rare Error

*Many factors, all necessary and only jointly sufficient, are required
to push a system over the edge of breakdown.*
– Sidney Dekker

The way an organization handles rare errors is a litmus test for how well leaders understand behavior as it relates to safety. Rare errors are often approached as individual or group failures, when in fact they are most often a failure of leadership.

There are usually two causes of rare errors. They are caused by inadequate training or by an environment that produces too few reinforcers to keep the employee focused on the task. By *focused* we mean that the performer is immune to elements in the environment that would tend to distract him.

All too often a safety investigation concludes that "performer error" was the cause of an accident when the real cause was a failure of training. This may seem to contradict what we have said earlier, that training is an antecedent and as such does not lead to permanent behavior change. It is not a contradiction for the simple reason that in order to execute a process or procedure safely, one must know how to do a task safely and have the motivation to perform it exactly as trained. Determining whether a safety problem is the result of inadequate training or ineffective consequences is the first line of analysis in an effective safety system.

Fluency Training And The Rare Error

Remember that *behavior fluency* is defined as automatic, non-hesitant responding. Its benefits are long-term retention of skills and knowledge, improved attention span, resistance to distraction, endurance, and application of the knowledge and skill to novel situations. The advantages of fluency to safety are obvious; however, safety training rarely reaches fluency and many organizations depend on long periods of on-the-job experience to reach it. The problem is that during the long period where habits are not at the fluency level, exposure is unnecessarily high.

The investment in training to fluency has a very large payback when all the costs of accident and injury are factored into the safety equation. As T.C. Cumming, a former Navy Seal says, "The more you sweat in times of peace, the less you bleed in times of war." He is referring to the intense training that Navy Seals undergo in order to perform flawlessly on their missions. Many accidents and injuries are due to inadequate training in that fluency was not demonstrated at the point of training. By not training to fluency, we save a penny that will later cost us a dollar . . . or much more.

As noted earlier, the way to determine whether an accident was caused by inadequate training or by workplace consequences is to use Mager and Pipe's "Can't Do/Won't Do" Test.

Can't-Do problems are training problems; Won't-Do problems are behavior consequence problems. The test for a Can't-Do problem is easy. If the performer can't do what was required if his/her life depended on it, it is a Can't Do. If the performer could do it if his/her life depended on it, but didn't, then it is a Won't-Do problem related to consequences.

In February 2009 a Colgan Air plane crashed into the suburbs of Buffalo, New York, killing all but one of the passengers and two people on the ground. The NTSB investigation concluded that the primary cause was pilot error. According to the report, the Captain did not notice that the plane's speed was dropping dangerously low and when cockpit warnings indicated the plane was about to stall, instead of pushing the stick forward to increase speed, he pulled it backwards multiple times.

A Colgan spokesperson said, "But they knew what to do in the situation they faced that night, had repeatedly demonstrated they knew what to do, and yet did not do it. We cannot speculate on why they did not use their training in dealing with the situation they faced."

Based on the Can't Do/Won't Do criteria, what do you think the primary problem was?

While the pilot clearly made an error, the focus of the investigation should be on *why* he made the error. The conclusion that a crash was due to pilot error tends to close the case, but it rarely gets to the real behavioral causes.

In spite of the company spokesperson's assertion to the contrary, the primary cause for the pilot's error was his training. While there may have been other contributing factors such as lack of sleep, it is clear that his training did not prepare him for responding properly to the conditions that he faced that night, which points to a failure of leadership.

It is always amazing to us that professional golfers are distracted by the slightest noise from the crowd or a passing plane. This is in reality a training deficiency. They could learn to perform under noisy conditions and it would benefit them because you never know when the next noise will occur. However, because there is a tradition of "quiet" by spectators and other golfers, the pros don't have to learn to do so.

Think of college or professional basketball players shooting free throws. The crowds cheer wildly and the fans try to distract the players by waving their hands, banners, and signs while yelling and screaming, all in full view and hearing distance of the players. Since the players have been trained under those conditions, they perform well under them.

Ron White, the 2010 U. S. Memory Champion, who memorized a shuffled deck of cards in one minute and 27 seconds, practiced for the competition by memorizing decks of cards under water. His trainer, T.C. Cummings, suggested that method as a way to learn under pressure and amid distraction. Most organizations are not prepared to train employees to that extent but clearly more of that kind of training would have saved the lives of passengers on the Buffalo flight. We suggest there are some life-critical procedures in many jobs that are worthy of this level of fluency training.

Although we are not sure, it is unlikely that the Colgan pilot had ever faced the particular situation that occurred on that night in February. He might have faced it in simulator training, but how many times? It was a very rare error and the question arises as to his history of reinforcement related to the desired behavior. It is clear that he was not fluent in the correct procedure, or at least unable to perform it automatically when under extreme pressure.

Ultimately, this disaster should not be looked at as pilot error; rather, it was a failure of leadership. It involved company policies, procedures, scheduling, and training, none of which was under the control of the pilot. Since he couldn't do the required behavior when his life depended on it, it was not a motivational (consequence) problem; it was clearly a training/fluency problem.

That said, to retrain employees when the problem is really about consequences in the workplace is a waste of time and money. Training or retraining will not solve a consequence problem. On the other hand, to treat a problem as one of consequences (motivational) when it is really a training problem can be disastrous.

The Role Of Consequences In Rare Error
Consider the following cases:

In the nuclear power business, going to sleep on the job is a serious offense. In 1987 the Peach Bottom nuclear plant was shut down when operators were caught sleeping. The problem of sleeping on the job is almost universally treated as a personal problem, and firing the offender is a common reaction. The trouble is that firing, suspending, or otherwise punishing the operators has not solved this problem. It still occurs every year. While there are certainly cases where an employee's behavior of drinking too much the night before or getting little sleep are the main factors, there are more cases where the problem is not personal habits so much as the way consequences are arranged for the work. Many attempts to eliminate this problem have not worked. Not surprisingly, over 700,000 hits on the Internet pertain to situations that involve sleeping on the job.

Although many strategies have been tried such as changing shift arrangements, giving multiple breaks, and training, one area that is seldom investigated for a possible solution is consequences for work performance. When a group of nuclear power plant managers complained of employees sleeping on the job, we asked them, "Has anyone here ever considered the fact that you have created a job that puts people to sleep?" Of course they hadn't. Most of their attempts involved changing antecedent conditions rather than consequences. They lectured, threatened, and lectured again, all to no avail.

—\m/—

We worked with a major automobile manufacturer in the days before bar coding. The plant had a spot on the production line where the frame and the chassis came together. The frame was assembled in another part of the plant and a conveyer moved the frame around the ceiling to the final assembly line where it was welded onto the chassis. Since they made a two-door car, a four-door car and a pickup truck on the same line, the sequencing of the two units was critical. The people on the line made a new car every 56 seconds, so when an order was sequenced improperly, the error needed to be caught quickly. If not caught quickly, this caused a serious problem as many units would be out of phase, which could result in a plant shutdown for up to a day. One of the plants we worked in rated the cost of the problem at $18 million annually and that was 20 years ago!

Because this was a costly problem, it made sense for the company to hire someone to stand at the mating location and check numbers on the frame and chassis to make sure that they matched. Since the error occurred very rarely, the employee always missed it at some point and they always fired him when he did. Interestingly, the replacement employee did great—maybe for a couple of years. However, eventually he too missed one and was fired. This had gone on for a number of years and was eventually solved by bar coding and a scanner.

—⁂—

We were conducting a seminar in a chemical plant in Texas. At the end of the first day of class, one of the participants asked if Aubrey had ever seen one of their control rooms. He responded that he had not but would love to see one and was subsequently invited to visit the control room and have dinner with the team.

When Aubrey arrived, he saw a control room with instruments that covered a wall about 40 to 50 feet long and a control station in the middle of the room. Behind the control station was a kitchen that any cook would covet, equipped with every modern appliance available.

Not only did the group have a sit-down meal, but Jake, one of the technicians, had made his famous blueberry cobbler and homemade ice cream for dessert. Just as they were finishing the meal and in the midst of "shooting the bull," an alarm sounded. Not one person at the table moved. Even though Aubrey is hard of hearing, he heard it and knew that surely they could hear it even above the *laughter and conversation. After several seconds of no response to the alarm, he asked, "Does that alarm mean anything?" One of the men jumped up and said as he left the room, "That thing has been going off all day. The alarm is not working right. We've called mainte-nance twice and they haven't been here yet."*

———

The performance problems exhibited in all three of the previous situations are caused by extinction. The positive reinforcer for the employees in each condition is "finding an error." Given that there are almost no errors to find, the behavior of "looking" undergoes extinction.

In the case of the auto plant, the employee could look for a mismatch of the frame and the chassis for 75 to 100,000 times and never see an error. In the chemical plant control room, extinc-tion occurred after only a few "looks" at the alarm. In the nuclear plant, the job remained the same night after night. There was no variation and all readings remained in the normal range. Looking and checking produced no variation—exactly what management wanted. Unfortunately, from a behavioral perspective, the job pro-duced extinction of the checking and monitoring behaviors.

Repetition does not make a job boring. Jobs that are consid-ered dull, boring, or ones that put people to sleep are most often the result of poor job design, not personal behavior. A boring job is one in which the reinforcement level is too low to keep the per-former's attention. It has little to do with repetitious behavior; it has everything to do with the reinforcement produced by the rep-etition. Go to Las Vegas or Atlantic City and watch people playing the slots. They will do this for hours on end and they pay to do it even though it involves the repetitive task of pulling a lever or pushing a button. It is not considered boring because the machine has been programmed to produce just enough reinforcement to keep the customer doing the same thing over and over and over. Repetition without reinforcement will cause even the most vigilant performer to eventually lose his/her motivation. We hope the reader realizes the folly of blaming the performer for loss of concentration or attention when the problem comes from the lack of adequate reinforcement produced by the system or by management behavior.

Extinction: The Unseen Obstacle In Safety

Extinction increases variation in behavior. Positive reinforcement can cause a person to respond in a precise way with very little variation for a very long period. Although many factors affect the rate of extinction, in general the rate is related to the number of reinforcers received for the behavior. Behaviors that receive few reinforcers in training and few on the job are subject to the effects of extinction in a matter of days or even hours. The following graph demonstrates that behaviors that receive fewer reinforcers extinguish more quickly.

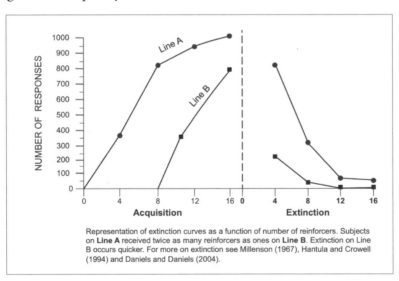

Representation of extinction curves as a function of number of reinforcers. Subjects on **Line A** received twice as many reinforcers as ones on **Line B**. Extinction on Line B occurs quicker. For more on extinction see Millenson (1967), Hantula and Crowell (1994) and Daniels and Daniels (2004).

Increased Risk Of Catastrophic Failure

An irony in safety is that many organizations with exemplary safety records have a higher risk of a catastrophic failure than do organizations that are actively working to improve their records. Complacency can accompany high safety performance and result in few reinforcers for monitoring behavior. Therefore, when an incident does occur, it has the potential to be catastrophic; no one will be looking because looking will likely have undergone extinction.

An early sign that extinction is occurring is that small variances in usual performance occur, but are not thought to be significant. These variances may take the form of misreading a dial or gage, not paying attention to an unusual noise, being preoccupied with some off-task situation, or failing to make a periodic check of equipment. In time, these variances will occur much more frequently and will catch everyone off guard when a major safety event occurs.

Job focus is determined by the frequency of reinforcers received. When the majority of the reinforcers available in a job setting come from doing the work, employees are highly focused on the job. When few reinforcers are produced by focusing on the work, it takes very little non-work activity to distract any performer—a sign of impending extinction.

From a leadership perspective, small variances in safety performance always demand attention since they may be an indication of extinction, the unseen obstacle in a high-performing safety culture.

No Harm; No Foul

In sports there is something called, "no harm; no foul." It means that someone did something that was against the rules, but since the outcome was not affected, no foul is called. Many people have used this approach in safety. If someone opens the wrong valve and water comes out, it will be viewed differently than if steam came out and burned someone. From a behavioral perspective, the behavior was the same—opening the wrong valve. Therefore, the consequences to the performer should be the same.

While at first glance this may seem neither correct or just, it is. The reason it doesn't seem right to most people is that the consequences for such a behavior are usually negative and it seems that the consequences for the employee whose behavior injured someone should be more negative than the one whose behavior didn't hurt anyone. The correct response in both cases is to ensure that safe behavior is fluent either through fluency training or on-the-job habit development so that the employees will be able to perform correctly in the future.

The Matching Law

Although many variables affect choice behavior, the matching law[1] stated generally, predicts that behavior will go to the most reinforcing element in the environment. When you observe employees in the workplace, you will see what is and is not reinforcing. Whatever people are doing pinpoints where the reinforcement is occurring. If employees are focused on the job, reinforcement is occurring for those work behaviors. If people are chatting among themselves rather than attending to work, you know that chatting is more reinforcing than working. If people are looking around, talking on their cell phones, reading a paper, or just daydreaming, you know that insufficient reinforcement is taking place to keep people focused on their tasks.

$$R = k\frac{r}{r + r_e}$$

R = Rate of approach of the target behavior to the asymptote
k = Asymptote (highest level for the target behavior)
r = Reinforcement for the target behavior R
r_e = Reinforcement for all other behaviors

This knowledge is highly diagnostic for safety leaders as they can see potential problems that relate not only to safety but to areas such as productivity and quality. When employees are not on task, there is a problem with the rate of reinforcement available for the performer on that task. When you understand this fact, it is easy to see that punishing the performer will not solve the problem.

Inattention, a frequent cause of safety problems, is a reaction to very low rates of reinforcement either from the task itself or from supervisors and managers for task performance. Unless actions are taken to increase reinforcement for the task, attention to the task will decrease and eventually extinction will occur. This occurs more often in monitoring tasks than most managers realize. Even though they are "on the job" they do not see what is happening as very little reinforcement is coming from the task and more is available for alternative behavior—like sleeping or daydreaming.

One of the first signs that extinction is occurring is when a seasoned and steady performer shows even a small variation in performance. The graph below is an example of early signs of extinction. Even slight changes deserve attention. While it is not necessary to respond to every variation in task performance, it should be observed and noted to see if it increases or decreases over time. In jobs like security, quality inspectors, airport screeners, safety observers, maintenance inspectors, and any job requiring checking or monitoring computer screens, equipment, or behavior, small changes are often ignored. When you understand that a small variation can indicate the beginning of the extinction of monitoring behavior, the problem should always be addressed to determine if it has a behavioral cause. If it has a behavioral cause, the solution is not to be found in applying negative consequences to the performer but in changing the consequences produced by the process, equipment, or management behavior.

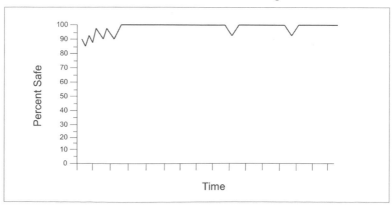

Wear Face Shield When Grinding

Methods For Protecting Against: The Rare Error

Having encountered all of the problems above, we have also been able to prescribe solutions based on the knowledge of extinction.

In the case of the performer in the automobile manufacturing plant, the first task was to determine the employee's reinforcer. His job was to find the error. Unfortunately for him it was possible that he could look for a mismatch (his positive reinforcer) for over a year before one occurred. Since looking at 75 to 100,000 cars produced no errors (no reinforcement) it would be accidental if he caught one when it occurred. Extinction of monitoring behavior would be complete after that occurrence.

We went to the performer and told him that we were going to purposefully change some of the numbers so he would have more to catch. Since we knew how many we put in the process, we were able to give him a score like any production employee. If we put in 100 errors and he caught 90, his score would be 90 percent accuracy for the day. He was told that before he stopped the line he was to check with the supervisor to see if the error was real or just one that we put in the system. We increased the number of errors we put in until he caught them all. Once his performance stabilized over a couple of weeks, the number we put in the process was carefully reduced until we made the input only occasionally.

The change in the man's interest for this job changed overnight. He was excited to see the score every morning. Please note that this was not used to catch the performer in an error but to help him attain a fluent performance that would last for long periods without the input errors. His performance determined what we needed to change to help him. His behavior was a function of the level of reinforcement that we put into the process.

We have used a similar technique in a wide range of jobs requiring checks of equipment or processes. Because of the high levels of quality and safety in some organizations, nothing changes over a very large number of checks or observations. Therefore some method of inducing changes (adding false positives) in the system are needed to sustain high levels of performance.

TSA screeners at airports look at large numbers of briefcases, suitcases, and so on without finding suspicious items. Aubrey carried what he thought was a legal knife in his suitcase only to find out after over 100 inspections that it was not legal. The solution is for TSA to put more items in the system and reinforce improvement in catching them.

Many performers go through safety checks and inspections overlooking errors in plain sight. This is not because they are lazy or careless but because the environment they work in has insufficient reinforcers to maintain high levels of alertness. It is the leader's responsibility to engineer those environments to produce effective levels of reinforcement. The employee's behavior is a reflection of how well the leader is doing his/her job.

Knowing the science of behavior in detail produces solutions to many problems that have plagued organizations for a very long time. Granted, common sense solutions work sometimes, but often they don't. Conversely, the laws of behavior never change. Once you know them you are able to solve problems in a new and ever-changing workplace.

---

[1]Herrnstein, R. J. (1997) *The Matching Law: Papers in Psychology and Economics.* Harvard University Press. Cambridge, MA.

CHAPTER 13

Effective Safety Leadership

Maintaining a Safe Physical Environment

The 2009 BP audit found that Transocean had left
390 maintenance jobs undone,
requiring more than 3,500 hours of work.
The BP audit also referred to the
amount of deferred work as "excessive."
– Ian Urbina, *New York Times*

Creating and maintaining a safe workplace is the ultimate responsibility for leaders in safety. In traditional occupational health and safety terms, this is about controlling the exposure to hazards. The hierarchy of controls provides a way to prioritize control methods.[1] Elimination (removing the hazard) and substitution (replacing hazardous products or procedures) are at the top of the hierarchy because they provide the best protection. Engineering controls are next in the hierarchy and involve trying to remove a hazard or put a barrier between the workers and the hazard. Administrative controls (implementing training procedures and policies) and personal protective equipment are considered the least effective, but necessary, where hazards cannot be controlled via other means.

Good safety leaders pursue controls at the top of the hierarchy (elimination and substitution) first and work down only when necessary. Completely ridding the work environment of hazards is obviously the best solution. Controls on the lower part of the hierarchy are less effective precisely because, rather than removing the hazard, they include changing the behavior of people in the workplace to avoid or minimize the hazard. Behavioral solutions are more difficult to sustain than physical solutions.

While ideally, leaders will design safe workplaces from the outset, most have inherited a physical environment and their challenge is to make it as safe as possible. This work is done largely through hazard identification and mitigation systems. Every organization has such a system, and they range from very informal to formal. Exemplars in safety have well-designed and well-executed systems. Some organizations have well-designed systems but have difficulty in execution. Some organizations have informal and sub-optimal systems. Wherever your organization falls on this continuum, our purpose is to discuss hazard identification and mitigation from a behavioral perspective. How can you encourage all the behaviors necessary to make such a system work effectively? We start with a discussion of one of the lesser-discussed benefits of mitigating hazards (the obvious benefit is a safer workplace) and then share some best practices that will maximize the impact of your system.

Hazards, Reciprocity, And Engagement

All human relationships rely on reciprocity to function effectively. Anthropologists and sociologists have documented the fact that reciprocity is a critical component of human interactions across cultures and over time. Reciprocity is considered one of the basic human norms.[2] In general, we are willing to put forth effort (to engage in behaviors) for others or for the greater good when we see others are also putting forth effort. If we see or believe that others are not putting forth effort, we are less likely to do so ourselves. Everyone has had the experience of working with a group (perhaps a fund-raising committee, children's sporting event, or special project at work) where everyone does not "pull their weight." Often one person or sub-group does a disproportionate share of the work, or not everyone completes his or her assigned tasks. This kind of inequity breeds discontent and when the activity is voluntary, often leads to withdrawal or reduced effort by those who were previously doing their fair share or more. While many people argue that safety is not voluntary, many components of safety are voluntary.

The components that move an organization toward being a high-performing safety organization are often voluntary, at least initially. BBS provides a perfect example. In order for BBS to be effective, frontline employees must be engaged in the process. A significant proportion of them must be willing to do observations, gather and analyze data, provide feedback to peers, and plan and execute celebrations around goal achievement. In our experience, the willingness of the frontline is directly proportional with the extent to which they see management actions in maintaining safe working conditions. It is basic reciprocity: if you do your part, I'll do mine.

While management's role in safety is bigger than just hazard mitigation, because hazards are tangible, they provide a visible way for management to demonstrate commitment to safety. We have found that organizations that deal with hazards effectively have an hourly workforce that is more engaged. When asked, they are willing to participate more fully in safety; to go beyond compliance.

The key to making the rule of reciprocity work for safety is being the one to "go first." After all, going first is the fundamental definition of a leader. (Webster's definition of *leader* is "something that leads.") When managers are willing to take the first step and change their behavior as it relates to safety, frontline employees will follow.

Critical Components Of A Hazard Process

We have seen a wide range of hazard identification and remediation processes. Most organizations have formal ways of identifying hazards including regularly scheduled inspections, work observation, surveys, and hazard discussions in safety meetings. These catch many hazards but just-in-time hazard identification processes are required as well so that employees can report hazards as they see them or as they arise, without delay or without having to wait for a formal inspection or the next safety meeting. In our experience, these just-in-time hazard reports most often take the form of a passing comment to supervisors or safety personnel. An operator might catch a supervisor as she is walking through the area and mention a leaking valve or a worn-out tool. Well-intending supervisors may or may not remember these informal reports and too many of the reports are lost. Hence, relying on verbal communication of hazards is not a good idea. A better process is required.

We believe all hazard identification and remediation processes should have three goals: (1) resolve hazards as efficiently as possible to create a safer workplace (2) increase the probability of reporting hazards, and (3) encourage employee commitment and enthusiasm around safety by taking action on formal and informal identification of hazards.

Taking a behavioral lens to this issue, we have come up with some critical components for hazard identification and mitigation. Please note: we are behavior specialists, not safety professionals. Our purpose is to evaluate processes from a behavioral perspective and provide suggestions for improving the behavioral components of hazard identification and mitigation.

MAKE REPORTING EASY

Most hazard identification systems involve too much response cost. Too much effort or hassle is associated with reporting a hazard. If an hourly employee has to fill out even a little paperwork to report a hazard, the probability of reporting goes down. For most people paperwork is a NIC that discourages reporting. It may seem crazy that an employee won't take a few minutes to report something that could cause injury, but if you understand behavior scientifically it is perfectly predictable. As noted previously, many organizations' informal reporting systems involve telling a supervisor or safety person in passing about a hazard. This information too often gets forgotten, leaving employees feeling as though no one is listening to them and that their concerns aren't respected. The key is to create a system that is easy for the person reporting, but captures the information every time. Ensuring it gets captured helps ensure it gets fixed which is the ultimate reinforcer for reporting (albeit a Future and Uncertain reinforcer). Seeing someone write down your concern and ask questions about the details is a PIC for most people.

Best practice: let those reporting hazards do so verbally and have someone else do the paperwork/enter it into the system.

MAKE ONE PERSON ACCOUNTABLE

If everyone is accountable, then no one is accountable.

Best practice: assign one person to be responsible for the hazard process. This person's responsibilities should include entering the hazards into the tracking system, monitoring the hazards over time, and shepherding them through to resolution. While a wide variety of people will both report hazards and address them, one entity should be responsible for making sure it happens. Finally, make sure everyone knows who the go-to person is.

TRANSPARENT PRIORITIES

Since resources are always limited, resolution of hazards must be prioritized. This is most often done through a maintenance work-order system and capital-request process. Safety work orders and capital requests are usually given priority over non-safety ones, but are still graded according to pre-established criteria. Sometimes disagreements arise around the priority given to particular hazards. Someone who has to work around a hazard every day will see it as a high priority, but in the broader context of other hazards it may be lower. When the prioritization is determined without input or even feedback to those who report the hazard, resentment builds.

Best practice: make the prioritization criteria public and include a feedback loop back to the person who reported the hazard informing them of the priority and the reason for the priority assigned, particularly if a hazard gets a low priority or is removed from the hazard list altogether.

PUBLICLY DISPLAY THE HAZARD LIST

Accountability increases when all employees can see which hazards have been reported and whether or not they have been resolved. Public displays generate social consequences from a variety of sources. First, the public list serves as a "to do" list for those responsible and may provide a source of reinforcement when items are checked off (much the same way checking off items on our personal to-do lists can be reinforcing). Second, employees who see that hazards on the list are being addressed are more likely to reinforce those who are responsible for the improvements. Third, when people see items that have not been resolved they are likely to put pressure on the appropriate parties, thus providing negative reinforcement which can spur action.

Another advantage of the public display is that senior management is more likely to be aware of the hazards and may be more likely to aid in resolution. More than once we have seen senior managers look at such public lists in passing and immediately do something about a hazard they see listed. Without public lists senior managers often do not know about many of the hazards.

Many organizations make their prioritized list of hazards available to all through company intranets and therefore believe they have a "public display." Unless your organization has hourly employees who fluently and frequently access your intranet, this is not sufficient.

Best practice: post the hazard list (including date reported, priority, and projected date of resolution) on bulletin boards in common areas.

SEEK INPUT ON RESOLUTION

Two objectives for resolving hazards have been reported: (1) to make the workplace safer, and (2) to increase enthusiasm and buy-in from the hourly population. Thus, to think any fix is a good fix is a mistake. If a hazard is not addressed according to the performer's expectation ("That's not the fix I had in mind!") then it may make the workplace safer, but it won't contribute to enthusiasm and buy-in. Why not get both?

Best practice: either involve the person reporting the hazard in the resolution planning or if others are planning the resolution, take the potential resolution to the individual who reported it and involve them in selection of the final solution.

INCLUDE FINANCIAL DECISION MAKERS

While much of the hazard process need not involve higher-level managers, the most successful systems include a high-level decision maker at critical points. Hazards will often be resolved more quickly, including lower priority hazards, when a plant manager or operations manager is involved. Warning: do not let the involvement of higher management in the decision process slow down the process of resolving the hazard.

Best practice: involve senior management.

CLOSE THE FEEDBACK LOOP FACE TO FACE

It is common to hear frontline employees complain that they reported a hazard and never heard about it again. This happens for at least three reasons:

1. The hazard never got formally reported.

2. It was given a lower priority or was taken off the list completely.

3. There is a delay in resolution due to funding, ordering of parts, et cetera.

Whatever the reason, if a person reports a hazard and then doesn't hear anything about it for weeks or months, they have just experienced extinction and are less likely to report hazards in the future. Individuals with this experience are also less likely to be engaged and enthusiastic about safety.

Best practice: after creating a feedback loop to keep employees who report hazards informed of the progress toward resolution, close the loop by verbally following up and having those who reported the hazard sign off that the hazard has been resolved. (Public displays of the hazard list accomplish this to some extent but it is usually not sufficient.) This extra step is usually positively reinforcing to the performer and serves as a final point to catch any problems or disagreements. One great way of doing this is through regular, hazard-status updates as part of existing meetings (such as start-up, shift change, or toolbox meetings). (Learn more at www.aubrey-danielsblog.com, "Unions and Performance Feedback.")

PUBLICLY ACKNOWLEDGE HAZARD RESOLUTION

For one of our clients, the final step is having those who reported the hazard publicly acknowledge that the hazard has been addressed. This serves as another form of feedback to all performers that hazards are being resolved. It also serves to decrease the likelihood of complaining that the hazard has not been addressed, or complaining generally about hazards not being addressed. When people state publicly that something good was done, they are less likely to contradict themselves later.

Best practice: have those reporting hazards (not the supervisor or manager) verbally acknowledge that the hazard has been addressed. Again, this is done most easily as part of regular hazard updates in existing daily or weekly meetings.

VISUAL DISPLAY OF PROGRESS

As noted earlier, employees who report hazards often do not get status updates and therefore believe their reports are being ignored. Companies who spend hundreds of thousands of dollars fixing hazards often get criticized by hourly employees for not doing enough. Perception is reality and perception is distorted by the fact that people are more likely to notice and remember hazards that don't get fixed because the hazard remains to be seen and experienced daily. Once a hazard is fixed there is often no reminder that it was ever there. It is difficult to notice the absence of a problem. Making the list of hazards public (as noted above) helps but be sure the list includes at least some of the hazards resolved, not just those yet-to-be resolved. Listing only outstanding hazards doesn't give credit for the hazards already addressed. Part of the reason to display this data is to demonstrate actions taken so that management is reinforced and hourly employees see progress.

Best practice: have graphic displays of hazard resolution in addition to public lists of hazards. Percent of hazards resolved, or average time to resolution are two possible types of data that could be graphed. Tables of numbers are harder to read and don't show trends over time. A picture is worth a thousand words. Such graphs help all employees see how many and how quickly hazards are being addressed.

Summary

The task of creating and maintaining a safe physical environment is an ongoing challenge, particularly for aging work sites. There is no such thing as perfection, and there will never be enough resources to remove all hazards. However, increasing communication through the methods outlined above goes a long way to creating a safer workplace and a feeling of cooperation and caring that generates willingness on the part of hourly workers to actively participate in safety endeavors.

Here is a checklist of the items discussed:

- Make reporting easy.
- Make one person accountable.
- Ensure transparent prioritizing.

- Publicly display the hazard list.

- Seek input on resolution.

- Include financial decision makers.

- Close the feedback loop.

- Publicly acknowledge hazard resolution.

- Visually display progress (with graphs, for example).

---

[1] The hierarchy of controls represents five methods of controlling hazards that are rank ordered according to effectiveness. NIOSH summarizes the hierarchy as follows: elimination (at the top), substitution, engineering controls, administrative controls, personal protective equipment (at the bottom).

[2] Cialdini, Robert B. (2009). *Influence: Science and Practice.* Boston: Pearson/Allyn and Bacon.

CHAPTER 14

Effective Safety Leadership
Encouraging Safe Behavior

*To be good is noble; but to show others how to
be good is nobler and no trouble.*
– Mark Twain

So far we have discussed the importance of developing good relationships, using science to better understand at-risk behavior, and maintaining a safe physical environment. Doing these well requires behavior change. Our focus thus far has been on the behaviors that need to change such as doing what you say you will do, conducting PIC/NIC Analyses of incidents, and creating systems to make reporting hazards easy. Next we will focus on how to increase and sustain those important leadership behaviors.

It's All Behavior
Because of the tremendous success of behavior-based safety, the term *behavior* is often misunderstood to represent only the behavior of frontline workers. The term calls to mind putting on hard hats, using three-point contact, and driving a safe distance from the vehicle in front of you. Frontline behavior is critical because the frontline is the most common point of injury. But a complete

behavior analysis shows that frontline behavior is largely a function of the upstream safe and at-risk **behaviors** of leaders.

Leader Behaviors: The Antecedents And Consequences For Frontline Behavior

Leaders create the physical and social work environments within which frontline work is done. Leaders decide which equipment to purchase, how to design custom equipment, and when or if to make changes or install upgrades. Leaders establish work flow, create work procedures, and decide on volume of work. Leaders are also responsible for the frequency and content of safety discussions, pressures and priorities experienced at the frontline, and the content and frequency of the day-to-day interactions between management and hourly personnel. All of these variables (and more), created and maintained by *management behavior*, establish the mix of antecedents and consequences that influence safe and at-risk behavior. A superficial understanding of behavior leads to the assumption that antecedents and consequences are delivered only in direct manager/employee interactions. The truth is that *anything* in the performer's environment can be a source of antecedents and consequences and therefore can influence behavior.

The diagram below illustrates the most common sources of antecedents and consequences for safe behavior. We will focus on

Sources of Antecedents and Consequences

consequences because of their disproportionate impact on behavior. Leaders have a hand in almost all of these sources. While not every leader can influence all of these sources, leadership as a whole can.

As noted earlier, leaders can change the physical environment through decisions regarding equipment purchases and hazard mitigation. Leaders create and change the organizational systems such as pay and promotion systems and contractor relationships. Leaders can alter the natural and work-process consequences through activities such as selection of PPE and SOP modifications. Leaders can encourage peer reinforcement through developing trusting work environments, and they can directly reinforce or punish safe or at-risk behavior. The optimum safety culture is one in which all these sources of antecedents and consequences work together to support safe behavior at all levels.

While everyone in the organization should participate in harnessing and directing all antecedents and consequences to support safety, leaders are in the position to orchestrate the activity. Leaders should engineer the consequences to the best of their ability to support safe behavior. This is no small task.

In the last chapter we talked about a process for addressing the physical environment through better hazard identification and mitigation processes. In this chapter we will talk about addressing contingencies (antecedents and consequences) that come from decisions, work process, people, natural/automatic sources, and organizational systems.

Decisions

Many organizational decisions impact safety. Indeed, often decisions on topics seemingly far removed from safety, influence safety. Thus it is our recommendation that safety be considered in *every* decision. This can be as simple as asking the question, "How will this impact safety?" during the decision-making process. However, those who answer the question need to understand behavior scientifically. For example, if as part of a quality improvement, a change is made to a process which makes it more difficult for performers to maintain a safe posture (or keep PPE on, or any other safe behavior) then this should be factored into the decision making. That is not to say the process change should not be adopted, but the effect on safe behavior must be recognized and a plan to deal with the effect on safety should be implemented. Like any other kind of system, a change in one area has ripple effects. The key to ensuring the ripples don't have an adverse impact on safety is in understanding the power of PICs and NICs on behavior. Too often, people anticipate the impact on safety but dismiss it by saying, "We will just train them to do it safely" or "We will put up a sign to remind them." Such antecedent solutions are unlikely to prevent the problem.

It may be asking too much for all those making decisions to have this level of understanding of behavior (although it would benefit the organization beyond just safety). The most practical approach may be to develop one or more individuals as the behavioral subject-matter expert(s).

The first step, however, is to ask the question, If safety is a priority then shouldn't it be deliberately considered as part of all decisions? Leaders should make it a habit to ask, "How will this decision impact safety?" and expect that they will get a behaviorally sound answer.

Work Process

The natural outcomes of following a work procedure are a source of consequences. One step of a process leads to another and so future steps become consequences for earlier steps. Analyzing steps of a process and looking for embedded consequences that reinforce at-risk behavior will alert organizations to potential problems. Ideally, the process would be modified to make safe behavior more likely. If not, the analysis serves as a prompt to make adjustments to other consequences to ensure the safe behavior still occurs despite work-process consequences that encourage at-risk behavior.

Natural/Automatic

Natural or automatic consequences are similar to work-process consequences in that they are consequences generated automatically by the behavior. That is, the act of engaging in the behavior automatically leads to the consequence. Putting on gloves reduces dexterity, no matter how good the gloves. Reduced dexterity is a natural consequence of putting on gloves. Getting a ladder takes more time than climbing on a nearby object (a chair, piece of equipment, and so on). Taking more time is a natural consequence of getting a ladder. Unfortunately, the natural consequences for most safe behaviors are negative. Many safe behaviors at the hourly level are often experienced as time-consuming, uncomfortable, and a hassle. Many safe behaviors at the supervisory and management level are experienced as "extra work" that is not quite as important as work that directly meets customer requirements and drives profit.

Can we do anything about natural consequences? Sometimes. Buying better gloves and making ladders more accessible will reduce the natural negative consequences, but not eliminate them. Every attempt should be made to modify natural consequences, because natural consequences are immediate and certain, therefore very powerful. Anything that can reduce or eliminate NICs for safe behavior is worth investigating. At supervisory levels, the natural consequence of "extra work" can be modified by changing what managers and executives emphasize and reinforce. If safety is only attended to when there are problems, then proactive safety activities will always feel like extra work. High-performing safety cultures are those in which proactive safe behaviors are not considered extra work, but essential work. That message is sustained and made meaningful in terms of actual behavior when supported with the proper consequences.

People-Generated Contingencies

People provide antecedents and consequences for safe and at-risk behavior every day. While we often assume the most powerful contingencies come from our boss, other people can and do sometimes provide even more powerful contingencies. Engaging everyone in the job of reinforcing safe behavior leads to the highest probability of success.

MANAGEMENT (THE BOSS)

What supervisors say and do as they interact with frontline employees will influence frontline behavior. What managers say and do as they interact with supervisors will influence supervisor behavior. In addition to the obvious things such as reminders to engage in specific safe behaviors, praise for safe behavior, or corrective feedback for at-risk behavior, it is also important to consider reinforcement available for behaviors that compete with safety. For example, if there is minimal reinforcement for critical safe behavior but a large amount of reinforcement for completing work on or before schedule, then workers will often work quickly, at the expense of safety. While working quickly and safely are often not mutually exclusive, performers often perceive that the two *are* mutually exclusive. They believe working the safe way will take more time and thus choose to focus on working quickly and meeting the schedule because their balance of consequences favors staying on schedule.

In addition to an immediate boss, this category also includes management at higher levels. While performers have less contact with higher-level leaders, what those leaders say and do can have great impact on behavior.

We know of a plant manager whose management style would have made a drill sergeant look meek. Being exposed briefly to the concept of positive reinforcement, he wrote in a shift log book, "Great work. Best production since I have been here." While this was intended to be a positive reinforcer for the crew (and was), it started unhealthy competition between crews during which they failed to pass on critical information to the oncoming shift and often left the production area in a physical condition that delayed efficient performance. When told about this unexpected response to his comment, the plant manager had no idea that a simple note could have such an impact.

We know of countless examples in which a brief, personal statement by an executive serves as a very powerful reinforcer that is repeated by performers for years. Of course negative examples are also common and are probably repeated more often than the positive ones.

PEERS

Peers significantly influence each other's behavior. Those who work in close proximity have many opportunities to deliver antecedents and consequences each day. Importantly, those consequences are more likely to be PICs and NICs than those provided by management. Unfortunately, this powerful source of consequences is not always working to support safety. While we all hope peers would prompt and reinforce each other for safe behavior, peers might also tease each other for complying with safety rules, model at-risk behaviors for new employees, or look the other way when a peer does something unsafe. Without specific training and encouragement, peers will often do nothing to encourage each other to behave more safely, report hazards, or generally be active participants in safety. Such behaviors must be shaped and nurtured. A BBS process based on positive reinforcement will do this. In implementing our BBS process we have seen repeatedly how the culture at the frontline changes from one of "keep to yourself and mind your own business" to one of "look out for each other." Again, this culture must be shaped carefully and deliberately. Without guidance in the use of feedback and positive reinforcement, peers often default to providing largely negative feedback (only talking to peers when they are doing something at-risk). This focus on the negative will move your organization further away from safety excellence.

As we discussed in an earlier chapter, one of the best ways to get peers engaged in more proactive safety behaviors is for management to improve the hazard identification and mitigation process. When hourly employees see management making additional efforts to maintain a safe physical environment, that action becomes a Motivating Operation (MO)[1] for others, making them much more willing to get involved in proactive safety activities.

DIRECT REPORTS

Antecedents and consequences don't just flow downhill. Direct reports are a significant source of antecedents and consequences for their bosses. How direct reports respond in safety meetings can influence the length and content of the meetings. Direct reports can punish their bosses' early attempts to reinforce. They can punish requests for their involvement. We have seen some cases where direct reports have successfully shaped a boss into leaving them alone almost all the time. Just as children provide powerful consequences for parents and athletes provide powerful consequences for coaches, direct reports provide powerful consequences for bosses. The question is, Are those consequences working for or against safety? We recommend complete transparency. Leaders who are working to change their own behaviors to better support safety should let direct reports know the behaviors they are working on and regularly ask for feedback on how they are doing. A reciprocal relationship of meaningful feedback about what works and what doesn't around promoting and supporting safety is most likely to yield effective and sustained leadership practices.

Organizational Systems-Generated Contingencies

Another source of antecedents and consequences are the organizational systems within which people work. These systems can support safety or they can provide barriers to safety. Since management is responsible for the creation and maintenance of these systems, it falls to them to ensure the systems support safety. Depending on your role in the organization, you may or may not have input/influence over these systems; however, a high-performing safety culture is one in which questions and concerns about the impact of systems on safety can be raised by anyone.

Most of the systems we review (below) exist for reasons outside of safety. If safety were the only concern, then it would be a relatively simple process to ensure these systems were all perfectly aligned to support safety. Organizations are filled with competing objectives. Thus, safety must compete to some extent with systems designed to drive productivity, quality, sales, et cetera. Those are important objectives and need contingencies to support them as well. However, in our experience, the impact of many of these systems on safety is not even considered. Bringing a behavioral lens to the analysis of systems impact will allow better decisions to be made about how to maximize their effectiveness, within and outside of safety.

Listed below are examples of systems which can impact safety. We provide examples of each. This list is not exhaustive as each organization has a unique set of systems. Our goal is to give you a flavor for typical systems and their potential impact on safety.

INCENTIVE SYSTEMS

Perhaps the best example of how an organizational system can negatively impact safety is an incentive system that focuses on productivity without a safety component. Workers who are paid more when they do more work will do more work at the expense of safety, quality, teamwork, and other organizational objectives. Most organizations learned this lesson the hard way and have worked to design incentives that include measures of all critical objectives, safety included. But the metric used is important. Too often the metric for safety is incident rate. The problems with incident rate as a measure of safety have been detailed elsewhere. The point here is that if you want employees to engage in safe behavior, then incentive systems should be assessed to see whether those systems encourage or discourage *safe behavior*.

HIRING AND PROMOTION

Hiring and especially promoting people who are known to be poor supporters of safety not only sends an undesirable message to employees, but doing so also disrupts efforts to shift the balance of consequences toward supporting safety. We have been highlighting the many ways that leaders influence safety. A leader who does not see safety as a priority is unlikely to engage in the considerable effort required to align all organizational antecedents and consequences to support safety. It is important to note that the higher the position of the non-supporter, the bigger the influence.

WORK PROCEDURES

Accident reports frequently list "failure to follow procedure" as a root cause for accidents. While BBS can help ensure that workers follow procedures, a first step should be to look at the procedure from a behavioral perspective. Some procedures are very difficult to follow even for a well-intentioned performer. Giving employees a cumbersome, time-consuming procedure and then placing them in a production-pressure environment is setting them up to fail. Obviously some procedures cannot be changed, but we have been pleasantly surprised at how often modifications can be made that make it easier for performers to do the right thing.

EQUIPMENT PURCHASE AND/OR DESIGN

Those whose job it is to purchase equipment are often charged to find a piece of equipment that does the job, fits the budget, and meets some basic safety criteria: it comes with safety guarding, for example. The consequences that machines provide for safe and at-risk behaviors are not usually considered. The fact that a machine has safety guards becomes irrelevant if the guards make work so time-consuming and the machine so cumbersome to use that operators jerry-rig ways around the guards. The same problem holds when engineers design or redesign equipment without asking for input from those who will operate it. Frequently, safety problems are built-in, literally. It is impossible for a purchasing agent or engineer to know all of the subtle but critical features of the equipment they design or purchase. It is best to involve those who are the end users so that safety problems can be avoided. We know of several cases, however, where input from operators is built into the engineering process, but safety problems are not avoided. As noted earlier, this may require input from a behavioral subject-matter expert. Thinking through how equipment will be used and looking for PICs and NICs is a difficult, but worthwhile, task.

SUPPLIER RELATIONSHIPS

As with large equipment, other supplies impact safety. Suppliers provide safety equipment such as gloves, goggles, cleaning equipment, component parts, and packaging supplies. Those who make the purchases or arrange supplier relationships often do not have safe behavior in mind because they are often far removed from the end users. For example, the ordering of PPE supplies is sometimes assigned to someone working in an office. Not being the end user of these items, such individuals may make buying decisions that decrease the probability of the PPE being used, such as ordering gloves that are sure to keep hands safe but reduce dexterity so dramatically they are rarely used. Another client example involved two suppliers, each providing the same components for building wood boxes. The boards from one supplier had a coating that caused them to stick together, so those using the boards to build boxes had to pry them apart—a time-consuming task that sometimes caused injuries. The other supplier used a different coating that didn't stick. A shift was made to obtain all supplies from the non-stick supplier, but only after the frontline users of the supplies got involved.

COMMUNICATION PROCESSES

High-performance safety cultures have frequent and effective methods for communicating about safety (and everything else). Safety should be discussed daily at all levels of the organization. Consistency in such communication requires established communication vehicles. E-mail and intranets are NOT the answer. A pattern we are seeing as organizations get leaner is too much reliance on e-mail and intranets as major communication vehicles. Face-to-face brief interactions that are either scheduled or that can and do happen spontaneously are required. Make sure such opportunities exist.

CONTRACTOR RELATIONSHIPS

Contractors can be a big part of the work environment. Contractors can serve as poor models and can put employees in harm's way through their at-risk behavior. They can also be positive models. While most organizations consider the safety of contractors during the hiring process, the metric used is usually incident rate. Our advice is to ask for and assess behavioral data as well.

QUALITY PROGRAMS

Quality improvement processes sometimes have unintended side effects. Changes designed to improve quality can change consequences for safe behaviors. Given the movement toward Lean work environments, assessing the impact of quality changes on safety is particularly important. Making an environment too Lean may present safety challenges yet unseen.

Making Consequence Management Manageable

Safety leadership is a big job. We have outlined all the sources of antecedents and consequences that drive at-risk and safe behavior and have suggested that it is your job to re-engineer these to better support safe behavior. So how do you eat this elephant?

One bite at a time.

In our BBS process we recommend that frontline observers create a long list of safe behaviors they need to work on, but focus on just a few behaviors at a time. We do this because it is in keeping with the known facts about behavior change. Behavior change is difficult. Focusing on a few behaviors at a time makes change more manageable, and it makes change less time-consuming so it is more likely to happen.

We make the same recommendation to leaders. Each leader could make many changes. Make a long list, but focus on a few behaviors at a time; we suggest no more than three. The key is to focus on the *critical* few. After creating a list, prioritize your action according to which behaviors will modify the most PICs and NICs for critical safe behaviors within your span of control.

Positive Accountability: The Glue Of A High-Performing Safety Culture

We have suggested many ways that leaders can change their behavior to better support safety and move their organizations closer to a high-performance safety culture. We outlined the following four categories of leader behavior.

1. Relationship development

2. Using science to understand at-risk behavior

3. Maintaining a safe physical environment

4. Creating systems that encourage safe behavior

We have also suggested that change is most successful when accomplished a little at a time. But regardless of whether you are trying to change one behavior or many, behavior change requires support. Proactive safety leadership behaviors must be reinforced if they are to persist. A positive accountability system that holds people accountable, not for incident rate, but for the behaviors that will drive lower incident rates, will ensure safety targets are met.

If you have an effective accountability system for production, quality, et cetera, then build behavioral measures of safety into those systems. As meaningful discussions about safety leadership behaviors are built into existing discussions of performance, proactive safety behaviors will be seen as a priority. When safety meetings and discussions are events separate from other business, the sense of safety being "extra work" is strengthened.

Accountability is about consequences. The default form of accountability is negative reinforcement. People engage in the behavior to avoid the bad consequences that might happen if they don't do the behaviors, such as the boss being upset, peers being disappointed, and/or looking like you aren't doing your job. Positive accountability is about ensuring that desired behavior receives positive reinforcement. No workplace will be completely devoid of negative reinforcement, nor should it be. However, the balance of behavioral consequences should be heavily skewed in favor of positive ones, because that is the only way to create a workplace where discretionary effort characterizes the entire culture.

Accountability requires measurement. The safety leadership behaviors we have been discussing are easily measured via simple checklists or logs. The measurement tool is less important than the discussion around the behaviors and ongoing assessment of impact and continuous adjustment. One of our colleagues, Dr. Joe Laipple, has created an efficient technique for helping management change any critical behavior. The process is called coaching for rapid change. Its simple design is a clever application of the science and what we know about the realities of today's work environments.

Few things are more reinforcing than seeing behavior change before your eyes. In coaching for rapid change, leaders can often see not only immediate behavior change but the reaction of the person whose behavior was affected. Much of the coaching is done in real time which enables PICs for all involved.

In coaching for rapid change, each leader meets with direct reports for very brief coaching sessions (2-3 minutes per person) each day. These "touchpoints" are focused on one-to-three behaviors that the coachee has agreed to work on and must relate to the business outcome, in this case improving safety. The leader asks pinpointed questions that allow quick assessment of two things: is the coachee doing the behavior they agreed to do and is that behavior having the desired impact? Frequent conversations establish the behaviors as priorities and allow real-time coaching for improving performance. The focus on impact ensures the behavior is driving business results and is worth the time and effort. The focus on impact also helps performers see the small improvements they might otherwise not attend to, thereby helping them come into contact with the natural reinforcers of their coaching efforts.

Many positive accountability methodologies are available. The point is to make sure you have an effective process that supports ongoing safe behaviors. The critical features of an effective positive accountability process are as follows:

1. Focus on behavior, not incident rate.

2. Ensure frequent conversations about target behaviors. Don't depend on e-mails or written reports.

3. Make sure the system provides a vehicle for delivering more immediate, certain, consequences for target behaviors. Quick success will breed success. If a behavior-change process leads to improvement within a few days or weeks, then you are more likely to continue and expand your effort.

Summary

We hope we have provided some good food for thought on how you might improve your personal safety leadership or the safety leadership of your organization as a whole. Remember, while there may be many things you want to change, success is most likely if you take small and steady steps. In the words of Indira Gandhi: "Have a bias toward action—let's see something happen now. You can break that plan into small steps and take the first step right away."

---

[1]MO or Motivating Operation is any environmental change that has two effects: one is that it increases the momentary effectiveness of a reinforcer: and two, it increases momentarily the behaviors that have produced that reinforcer in the past.

CHAPTER 15

Effective Safety Leadership
Safe For What?
Pulling It All Together

He that will not apply new remedies must expect new evils.
– Francis Bacon

We believe that to improve safety some fundamental changes are required in leadership practices. As we have pointed out, some of the practices currently used are ineffective and in some cases counterproductive. Other, high-impact practices (outlined in this last section of the book) have not yet been widely adopted. We understand the futility of advising leaders to simply do more when there is no more time left in each day. Our hope is that you can find ways to eliminate less-effective practices and replace them with more-effective ones. We understand that although safety may be your top priority, it is not your *only* priority.

What Is The Mission Of Safety?

Is the mission of a safety department to have no accidents or injuries? Many safety professionals act as though it is. Is it to create a safe work environment? The Steelworkers think it is.[1] The perfect way to have no accidents or injuries, and thereby have a completely safe workplace, would be to buy ergonomically designed chairs that employees sat in without moving all day. While this sounds ridiculous, the outcome of such a workplace would satisfy most organizational safety goals. In other words, work interferes with safety. It is not all that unusual to have departmental goals that compete with organizational goals. To have zero defects in an automobile plant, we could limit production to one car per day. That would allow adequate time for redundant testing of all parts and systems. For maintenance to eliminate machine failures, we could take machines down for maintenance every day and run them at slow speeds. To achieve zero accidents, employees could work very slowly, considering safety and/or reviewing safety procedures before they do anything at all.

Here are some questions to consider: Can production goals be met and safety goals not met? Can quality goals be met and safety goals not met? Can the same be said for maintenance, sales, and training? What organization exists where no organizational goals can be met if safety goals are not? Is there such an organization?

Production and service goals can be met in all organizations even when safety goals are not. Getting "stuff" out the door often pleases executives, shareholders, and Wall Street. Of course, getting the "stuff" out of the door in a productive, quality, and cost-effective way is why organizations exist.

Sadly, it is all too clear that employees can and will engage in unsafe behavior to get stuff out the door. Many pictures circulate on the Internet showing people engaging in outlandish behaviors in an attempt to get the job done. One shows an employee repairing an electrical circuit standing on a metal ladder in two feet of water. Another shows a large forklift hoisting a smaller forklift to reach a machine that needs to be moved.

In fact, there are no organizations we can think of where safety is *the* mission.

What then is the Mission of the Safety Department?

We think the above question is answered by asking another, "What is the mission of the organization?" The mission of a furniture company is to sell furniture. Style, comfort, quality, and price all play a role in a person's decision to buy furniture. No one we can think of buys a recliner because the company has a safe workplace. Consumers don't even know the safety records of these companies.

The mission of the safety department in a furniture company is to help sell furniture! To help sell furniture, safety must be fully integrated into the process of making furniture better, faster, and cheaper.

Years ago, Aubrey was asked by the owner of a textile company to talk to Nick, the president of his carpet division, about implementing a performance management system in his mills. After Aubrey had waited a full two hours, Nick burst into the room, plopped down in a chair at the other end of the conference table and before Aubrey could introduce himself, Nick pointed his finger in his direction and said in a gruff tone, "Can you help me make carpet?" If managers and executives like Nick do not see safety as helping them make carpet, or whatever their company's product or service may be, how vigorously will they attend to safety on a daily basis?

Safety does not, or should not, exist in a vacuum. Every organization exists to produce a product or service that consumers are willing to buy. If safety doesn't facilitate business, it runs against the natural flow of the organization and the safety department will always be in a position of having to push managers and supervisors to view safety as a number-one priority when in reality it is not.

The mission of the organization and that of the safety department should be one and the same. The mission of safety in any organization should be to assist the organization in doing what it does while doing it safely.

What is a High-Performance Safety Culture?

Would you approve of someone doing something that improves quality and productivity but puts employees at increased risk for health and safety problems? Of course you wouldn't! Is it equally unrealistic to expect people who are in charge of productivity and quality to support safety when to do so interferes significantly with productivity and quality? Some safety people feel justified in requesting actions that interfere with the business because they feel that moral and civil law are on their side. To be certain, nothing that exposes people to health and safety risks can be tolerated in today's business. But the business must go on, and go on safely. Is it possible to meet safety, quality, and production goals at the same time? You bet it is. Sometimes, in the heat of battle, it does not appear to be possible, but later the possibility and necessity to do so becomes clear.

An organization where the mission of one area interferes with that of another will never achieve its highest potential. Unfortunately, in most organizations, even those where talk of teamwork is high, you find that people, shifts, departments, and initiatives are not linked for success. When you find such a situation, you usually find weak, ineffective leaders.

Think of the money spent on repairing the damage from catastrophic events when a strict adherence to safety protocols would have prevented those events. The strict adherence might have caused production to be delayed, customers to be inconvenienced, or costs to be high, but the catastrophes brought about those results anyway. Only a poor manager would continue to run product with a known quality defect. It is also poor management to continue to run product while ignoring obvious deviations from safe workplace practices. (Learn more at www.aubreydanielsblog.com, "How the Mighty Fall.")

So how can all these objectives be reconciled? An apt analogy is to view organizational leaders as plate spinners in the circus.[2]

Production, quality, efficiency, and safety are all plates. Sometimes the plates wobble and require attention. The challenge is that paying too much attention to one plate often leads to more wobbling plates, and those plates that wobble too much could even crash to the ground. The skilled plate spinner (leader) can keep all the plates spinning effectively. The key is fluency.

Safe behaviors (at all levels) must be habitual. In other words, all employees need to do their jobs safely without having to think of safety. When you drive a car, you don't have to think about slowing down when you see the brake lights come on in the car in front of you. We know that sounds a bit heretical, but having to think about safety while engaged in an activity illustrates that the behavior is not at a fluent level. It also interferes with performing your job efficiently. An article by Beilock, Carr, MacMahon and Starkes[3] *When Paying Attention Becomes Counterproductive*, shows that while thinking about specific behavior or actions may facilitate the performance of novices, it actually interferes with the performance of experts. It is helpful for novices and those seeking to improve performance to focus attention on their actions, but not so with experts. Targeting high-impact safe behaviors at all levels and then working towards fluency enables safe, productive, quality work.

Of course safety is not static. There will always be new or modified safe behaviors to work on just as there will be new quality behaviors and new productivity behaviors. The point is when new behaviors are identified they should be targeted for habit development. Even leadership behaviors (such as those described in this book), hazard identification/inspection behaviors, and low-frequency safety behaviors can become habitual. (See *Removing Obstacles to Safety*[4] for further discussion of safe habit development.) Once safe behaviors are habitual they no longer interfere with production; in fact, they help production because they prevent the significant losses associated with accidents and incidents.

Years ago, before Dr. W. Edwards Deming arrived on the quality scene, when questioned about poor quality, someone would inevitably ask, "Do you want quality or production?" The answer, of course, was, "Both." The answer to the question "Which do you want production, quality and cost, or safety?" is "All of the above."

An organization cannot sustain profitability when compromises are made in any of these critical operational aspects. Only a poor leader accepts compromise in one or more of these categories. PICs will tempt the weak leader to take shortcuts, but the strong leader will find a way to do the right things right. That means making sure that safety is fully integrated into all aspects of an operation. A high-performance safety culture is one where every employee is focused on delivering a product or service in a productive, quality, cost-effective, and *safe* way.

---

[1]Frederick, J. (1999) Comprehensive Health and Safety vs. Behavior-Based Safety: The Steelworker Perspective on Behavioral Safety (Part 2). Remarks to the 1999 Behavioral- Safety Now Conference. Las Vegas, Nevada, October 6.

[2]Plate spinning is a circus act in which the performer spins multiple plates on poles without letting any of the plates fall off and break.

[3]Beilock, S.L., Carr, T.H., MacMahon, C. and Starkes, J.L.(2002) When Paying Attention Becomes Counterproductive: Impact of Divided Versus Skill-Focused Attention on Novice and Experienced Performance of Sensorimotor Skills. *Journal of Experimental Psychology: Applied*, Vol. 8 (1) pp. 6-16.

[4]Agnew, J.L. & Snyder, G. (2008). *Removing Obstacles to Safety: A Behavior-Based Approach*. Atlanta: Performance Management Publications.

SAFE BY ACCIDENT?

Appendix

Suggested Readings

Books

Agnew, Judy, and Snyder, Gail. *Removing Obstacles to Safety*. Atlanta: Performance Management Publications, 2000.

Bailey, Jon, and Burch, Mary. *How to Think Like a Behavior Analyst: Understanding the Science that Can Change Your Life*. New York: Taylor and Francis, 2006.

Cameron, Judy, and Pierce, W. David. *Rewards and Intrinsic Motivation: Resolving The Controversy*. Westport, CT: Greenwood Publishing Group, Inc., 2002.

Cooper, J. O., Heron, Timothy E., and Heward, William L. *Applied Behavior Analysis*. Columbus, OH: Merrill Publishing Company, 1987.

Daniels, Aubrey C. *Bringing Out The Best In People*. New York: McGrawHill, 2000.

Daniels, Aubrey C., and Daniels, James E. *Measure of a Leader*. New York: McGrawHill, 2006.

Daniels, Aubrey C. *Other People's Habits: How to Use Positive Reinforcement To Bring Out The Best in People Around You*. Atlanta: Performance Management Publications, 2007.

Daniels, Aubrey C. and Daniels, James E. *Performance Management: Changing Behavior That Drives Organizational Performance*. Atlanta: Performance Management Publications, 2006.

Dekker, S. *Just Culture: Balancing Safety and Accountability*. Burlington, VT: Ashgate Publishing Company, 2007.

Eisenberger, Robert. *Blue Monday: The Loss of The Work Ethic in America.* New York: Paragon House, 1989.

Ericsson, K. Anders, Charness, Neil, Hoffman, Robert R., and Feltovich, Paul J. *The Cambridge Handbook of Expertise and Expert Performance.* MA: Cambridge University Press, 2006.

Gawande, A. T*he Checklist Manifesto: How to Get Things Right.* New York: Metropolitan Books, 2010.

Gilbert, Thomas F. *Human Competence; Engineering Worthy Performance.* San Francisco: Pfeiffer, an imprint of Wiley, 2007.

Laipple, Joe. *Precision Selling: A Guide for Coaching Sales Professionals.* Atlanta: Performance Management Publications, 2006.

Latham, Glenn. *Power of Positive Parenting.* UT: P&T Ink, 1990.

Lattal, Alice Darnell, and Clark, Ralph W. *Ethics at Work.* Atlanta: Performance Management Publications, 2005.

Maloney, Michael. *Teach Your Children Well.* MA: QLC Educational Services, Cambridge Center of Behavioral Studies, 1998.

Millenson, J.R. *Principles of Behavior Analysis.* New York: MacMillan & Co., 1967.

Mlodinow, L. *The Drunkard's Walk: How Randomness Rules Our Lives.* New York: Pantheon Books, 2008.

O'Brien, Richard M., and Eds. Dickinson, Alyce, and Rosow, Michael. *Industrial Behavior Modification: A Learning Based Approach to Industrial Organizational Problems.* New York: Pergamon Press, 1982.

Pierce, W. David, and Epling, W. Frank. *Behavior Analysis and Learning.* NJ: Pearson Education, 1998.

Reason, J.T. *Managing the Risks of Organizational Accidents.* Burlington, VT: Ashgate Publishing Company, 1997.

Sidman, Murray. *Tactics of Scientific Research.* Boston: Authors Cooperative, 1960.

Sidman, Murray. *Coercion and Its Fallout.* Boston: Authors Co-operative, 1989.

Skinner, B. F. *About Behaviorism.* New York: Knopf, 1974.

Skinner, B. F. *Science and Human Behavior.* New York: The Free Press: MacMillan, 1974.

Skinner, B. F. *Technology of Teaching.* New York: AppletonCen-turyCrofts, 1968.

Articles

Binder, C. (1996). Behavioral fluency: Evolution of a new paradigm. *The Behavior Analyst*, 19, 163-197.

Catania, A. C., Matthews, B. A., and Shimoff, E. Instructed versus shaped human verbal behavior: Interactions with nonverbal responding. *Journal of the Experimental Analysis of Behavior* 38 (1982): 233248.

Eisenberger, R. Learned Industriousness. P*sychological Review* 99 (1992): 248267.

Ericsson, A. The Role of Deliberate Practice in the Acquisition of Expert Performance. *Psychological Review.* 3 (1993): 364.

Green, L., and Freed, D. E. Behavioral Economics. In W. O'-Donohue (Ed.), *Learning and Behavior Therapy* (1988): 274300. Needham Heights, MA: Allyn & Bacon.

Hantula, D. A. and Crowell, C.R. (1994) intermittent reinforcement and escalation processes in sequential decision making: A replication and theoretical analysis. *Journal of Organizational Behavior Management*, 14, 7-36.

Iwata, B. Negative Reinforcement in Applied Behavior Analysis: An emerging technology. *Journal of Applied Behavior Analysis* 20 (1987): 361378.

Johnson, K. R., & Layng, T. V. J. (1992). Breaking the structuralist barrier: Literacy and numeracy with fluency. *American Psychologist*, 47, 1475-1490.

McDowell, J. J. Matching Theory in Natural Human Environments. *The Behavior Analyst* 12 (1988):153166.

Nevin, J. A. Reinforcement Schedules and Response Strength. In M. D. Zeiler and P. Harzem (Eds.) *Advances in Analysis of Behaviour.* (Vol. 1) (1978). *Reinforcement and the organization of behaviour* (117158). New York: Wiley.

Skinner, B. F. What is the experimental analysis of behavior? *Journal of the Experimental Analysis of Behavior* 9 (1996): 213218.

Journals & Magazines

The Behavior Analyst. Association of Behavior Analysis International, 550 W. Centre Ave., Portage, MI 490245364 | Phone: (269) 4929310 | Fax: (269) 4929316 | Email: mail@abainternational.org.

Journal of Applied Behavior Analysis [JABA]. Department of Applied Behavioral Science, University of Kansas, Lawrence, KS 660452133.

Journal of Organizational Behavior Management [JOBM]. Philadelphia: Haworth Press, haworthpress@taylorandfrancis.com.

Performance Management Magazine Online. Atlanta, GA: Performance Management Publications, 678.904.6140, www.pmezine.com.

About the Authors

Dr. Judy Agnew is Senior Vice President of Safety Solutions at Aubrey Daniels International (ADI). With more than 19 years of consulting experience and a Ph.D. in Applied Behavior Analysis, Agnew partners with clients to create behavior-based interventions leading to optimal and sustainable organizational change. Agnew has worked in a variety of industries including oil and gas, mining, forest products, utilities, food and non-food manufacturing, distribution, assembly, insurance, banking, newspapers, and retail. In addition to her consulting, project management and instructional design work, Judy is recognized as a thought leader in the field of behavioral safety and performance management. Judy has presented at major safety conferences including the National Safety Council and Behavioral Safety Now as well as other key corporate conferences. She is the author of *Removing Obstacles to Safety* (with Gail Snyder). A native of Calgary, Canada, Judy earned her BA, MA, and Ph.D. in behavioral psychology from the University of Victoria. Judy currently lives in California with her husband and two children.

Dr. Aubrey C. Daniels, founder of ADI, has devoted more than 30 years to working with organizations of all types and sizes to apply the science of human behavior in their workplace. A passionate thought leader and an internationally recognized expert on management, leadership, and workplace issues, Daniels has been featured in USA

Today, The Wall Street Journal, The New York Times, The Washington Post, Fortune, CNN, and CNBC. Daniels is a member of the Board of Trustees of both Furman University and the Cambridge Center for Behavioral Studies. He is an Associate of Harvard University's John F. Kennedy School of Government, an adjunct faculty member of the College of Health Professions at the University of Florida, and a visiting professor at Florida State University. Daniels' most recent book is *Oops! 13 Management Practices that Waste Time and Money (and what to do instead)* and he is also the author of four best-selling books widely recognized as international management classics: *Bringing out the Best in People, Performance Management, Other People's Habits,* and *Measure of a Leader* (with James E. Daniels). He received his doctorate from the University of Florida, his undergraduate degree from Furman University, and has been honored by both universities as Alumnus of the Year. Daniels and his wife Rebecca reside in Atlanta and have two married daughters and three grandchildren.

About ADI

Regardless of your industry or expertise, one thing remains constant: People power your business. At Aubrey Daniels International (ADI), we work closely with the world's leading organizations to accelerate their business performance by accelerating the performance of the men and women whose efforts drive their success. We partner with our clients in a direct, practical and sustainable way to get results faster and to increase organizational agility in today's unforgiving environment.

Founded in 1978, and headquartered in Atlanta, GA, we work with such diverse clients as Aflac, Duke Energy, Lafarge, Malt-O-Meal, M&T Bank, Medco, NASA, Roche Labs, Sears, and Tecnatom to systematically shape discretionary effort—where people consistently choose to do more than the minimum required. Our work with clients turns their strategy into action. We accomplish this not by adding new initiatives to their list, but by helping them make choices that are grounded in an ethical approach to people and business, by increasing effective and timely decision-making, and by establishing a culture of respect for each person's contribution, regardless of rank.

Whether at an individual, departmental or organizational level, ADI provides tools and methodologies to help move people towards positive, results-driven accomplishments. ADI's products and services help anyone improve their business:

Assessments: scalable, scientific analyses of systems, processes, structures, and practices, and their impact on individual and organizational performance

Coaching for Impact: a behaviorally sound approach to coaching that maximizes individual contributions

Surveys: a complete suite of proprietary surveys to collect actionable feedback on individual and team performance, culture, safety, and other key drivers of business outcomes

Certification: ADI-endorsed mastery of client skills in the training, coaching, and implementation of our key products, processes, and/or technology

Seminars: a variety of engaging programs of practical tools and strategies for shaping individual and organizational success

Scorecards & Incentive Pay: an objective and results-focused alternative to traditional incentive pay systems

Behavior-Based Safety: a proactive and systematic process for managing safety that creates a culture of safe habits

Safety Leadership: a behavioral approach to creating an high-performance safety culture through leadership action

Expert Consulting: custom, hands-on direction and support from seasoned behavioral science professionals in the design and execution of business-critical strategies and tactics

Speakers: accredited and celebrated thought leaders who can deliver the messages your organization needs on topics such as sustaining your gains, accelerating performance, and bringing out the best in others

www.aubreydaniels.com
www.twitter.com/aubreydaniels
www.youtube.com/aubreydaniels

Register Your Book

Register your copy of *Safe by Accident?* and receive exclusive reader benefits. Visit the Web site below and click on the "Register Your Book" link at the top of the page. Registration is free.

www.pmanagementpubs.com

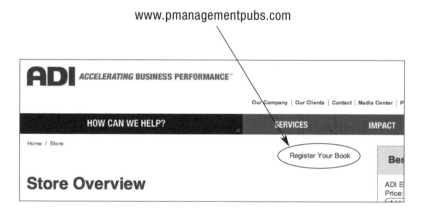

Performance Management Publications
Additional Resources

Removing Obstacles to Safety
Judy Agnew
Gail Snyder

Performance Management
(4th edition)
Aubrey C. Daniels
James E. Daniels

Oops! 13 Management Practices that Waste Time and Money
Aubrey C. Daniels

Other People's Habits
Aubrey C. Daniels

Measure of a Leader
Aubrey C. Daniels
James E. Daniels

A Good Day's Work
Alice Darnell Lattal
Ralph W. Clark

Bringing Out the Best in People
Aubrey C. Daniels

You Can't Apologize to a Dawg!
Tucker Childers

Precision Selling
Joseph S. Laipple

The Sin of Wages!
William B. Abernathy

For more titles and information call
1.800.223.6191
or visit our Web site
www.PManagementPubs.com

ADI Safety Leadership

In today's business climate, leaders face many challenges and pressures around production, schedule, quality, and keeping shareholders happy. While executives, managers, and supervisors know that creating a culture of safety is critical, they typically struggle with *how*. ADI's approach, grounded in proven behavioral technology, facilitates the development of a high–performance safety culture through leadership action. ADI works with clients to eliminate ineffective leadership practices and replace them with measurable, leader behaviors that positively impact safety. Working throughout the organization, ADI helps clarify roles and the behaviors necessary to create a culture of safety as well as develop the infrastructure to sustain those behaviors for the long run. Contact ADI to learn more about which safety solution will help your organization build and sustain a safety culture.

678.904.6140

info@aubreydaniels.com

www.aubreydaniels.com/improve-safety